passthecivilPE Guide Book

www.passthecivilPE.com

Exam topics covered include:

- ✓ Construction
- ✓ Transportation
- ✓ Structural
- ✓ Water Resources and Environmental
- ✓ Geotechnical

Updated for 2015 Specifications!

Now including the 8 NCEES distinct categories for breath exam topics!

Project Planning, Means & Methods, Soil Mechanics, Structural Mechanics, Hydraulics & Hydrology, Geometrics, Materials, and Site Development

4th Edition

COPYRIGHT INFORMATION

ISBN: 978-1-62141-945-7

TABLE OF CONTENTS

How to Pass the Civil Professional Engineering (P.E.) Exam

- After buying every guide, book, DVD, and online resource to use as reference and to help with studying, many test takers find that they still do not have enough diverse problems or solutions with thorough explanations to give them the confidence they need to pass the Civil Professional Engineering (P.E.) Exam.

- Because of all the frustrated would-be engineers and lack of available information, passthecivilPE makes its best attempt to provide the best possible problems and solutions to references during the exam.

- passthecivilPE has evaluated the most effective resources possible and considers it our obligation to lead others to them.

- This guide was developed because we know that practice is the most essential component to passing the Civil Professional Engineering (P.E.) Exam. Training with materials similar in format, timing, language, and style will increase your comfort level and prepare you for mastering the exam when it counts the most.

- This guide provides you with necessary information in the form of a combined practice exam and study guide that will give you confidence and prepare you for passing the Civil Professional Engineering (P.E.) Exam.

Introduction

This passthecivilPE guide was built to help would-be civil engineers pass the Civil Professional Engineering (P.E.) Exam by giving them the BEST resources possible, all in one place. This guide can be used during the exam to help tackle similar problems effectively and efficiently.

- Practice Problems and Solutions

  - Use the practice problems like a test; take the test a minimum of three times. Keep notes and track your progress.

  - First, go through the problems (without solutions) as if you are taking the test, timed and all, mark any references where you find equations or help with tabs or a sticky note. Don't look at the answers yet!

  - Next, see how you did and check the solutions.

  - Reread the Exam Advice. Use pointers such as skipping problems and marking references.

  - Retake the test, again. DO NOT look at solutions. Check your improvement. Also reread the Exam Advice. Use pointers such as back-calculating and circling or underlining problem statements prior to making any calculations.

  - Take timed tests to prepare you for exactly what to expect on test day so you are comfortable and confident in the exam room.

  - Write up detailed solutions on the pages, bind and tab them, and refer to the solutions during the exam.

- Process for Answering the Problem

  - At the beginning of each question, you are asked to quickly indicate several key points about the problem. This helps you SLOW DOWN and ask yourself if you understand what you are reading. There are also some key tips to help you SAVE TIME during the exam.

  - You will also need to reference the correct material - Is it a structural problem? Geotechnical? Always ask yourself what the problem type is and what it is asking for BEFORE doing any calculations or writing.

  - Use the exam booklet during the test to write down notes, cross out bad answers, etc. Once you have found the correct answer, circle it in the exam booklet - you can wait to fill in the correct bubbles on the answer sheet until the end of the exam; this should save you time during the exam.

  - Manage your time! See Exam Advice and General Advice for more about managing your time for each problem.

"QUICK TIPS"

These are in the solutions section and are to be used if you need a little push in the right direction.

These are to help prevent you from looking at the solution and keep trying to work the problem as if you were in a test situation – but remember, there are NO hints on the exam.

"THINGS TO REMEMBER"

These are in the solutions section and provide an additional resource related to the problem.

"THINGS TO THINK ABOUT"

These are in the solutions section and can be used to provide you with additional understanding of the individual topics.

© 2015 | passthecivilPE.com

7

General Advice

- Study using all your references in the same bag, suitcase, cardboard box, etc. that you are going to use during the test. Put the materials down on the floor next to you, or on the desk directly in front of you, since that's where they will likely be during the test.

- Study using a 0.5 lead or #2 pencil. The pencils they have for the test are not the best – get used to writing with the smaller diameter pencil lead so you're not breaking or running out of lead throughout the test.

- Study using the exact same Approved Calculator you will use on the exam. This will help you make efficient use of your calculator time.

- Bring an extra approved calculator, and leave in your bag just in case.

- Bring a square pillow for your seat at the exam site. The collapsible seats are uncomfortable and are too low in relation to the long folding tables. Not only does it lift you up several inches but it's comfortable!

- Bring a healthy, packable lunch. Eat it on your short break between tests by yourself somewhere quiet.

- Don't meet with others before the exam or at lunch time, unless you will be studying. Resist the urge to gossip with others, unless you think this will relieve some stress.

- Don't drink too much coffee, but do make sure to drink enough water on test day.

- Use the restroom, drink some water, and eat some solid food before the exam.

- Be prepared to sit and wait for others to file into the exam room and fill out paperwork for an hour or more after the doors to the exam site open before taking the first 4-hour test.

- Get to the exam site early, find a quiet area, and review your notes.

Exam Advice

- It is extremely important to consider your time management on the exam; if you can nail down your time on each problem, you will have maximized your chances of a higher score on the exam.

- Make a first pass through the test answering questions; this should take a couple hours, at least. Don't waste time scanning through the exam without answering questions, it just wastes time. Instead, read, take notes in the margins, circle or underline questions, and try to solve the problems you know on the first pass.

- You will have to practice doing a first pass before the test, and you have to follow the test advice below in order to maximize your time on each of the questions. Your first pass should nail the questions you absolutely know – but that means you have to READ the ones you don't absolutely know, which takes time – the trick is to read thoroughly and take notes in the margins on the problems you don't really know.

- Skip problems on your first pass if you have no idea what the question is asking. As a soon-to-be engineer, this will likely bug the snot out of you – don't let it rattle you. Move on and come back. Have you ever not been able to remember a word or phrase when you're talking to someone? And then twenty minutes later you suddenly remember?! This happens on tests all the time, especially with the added anxiousness of a timed national test!

- When you're lost in the problem, go to the margin and write down the type of problem, a few notes about what you're thinking the right path is, underline or circle key words and what the problem is asking for in the problem statement, and then MOVE TO THE NEXT PROBLEM! When you come back, you may have new insight...

- If you are going to skip a problem:

    - Read the answers to see if they help make the problem understandable.

    - Try to eliminate at least one answer, try for two...if you eliminate three then bubble in the last one and you're done!

    - Quickly write down a couple questions in the margin to help you when you come back.

- Write clearly and in steps, so when you don't get a good answer you can go back and check (see below).

- Circle answers in your exam booklet. Do not fill in answer sheet bubbles until you finish your first pass! At the end of the first pass, fill in your bubbles and then you will be able to easily find the problems you need to go back to by looking at the empty bubbles on the sheet.

- For each problem take note of, or quickly write down, the following:

    1. What type of problem is it? (5 seconds)
    2. What is the problem asking for? (10 seconds)
    3. Is there extraneous information? (10 seconds)
    4. What references or equations are needed? (10 seconds)
    5. Use the space on your sheet (if printed out) or scratch paper to find your solution.
    6. Circle the right answer (if printed out) or write down the correct letter (do not fill in the score sheet bubble yet).

- If you don't see your answer in the multiple choice questions, do these things, in this order:

    1. Quickly check to see you're using the correct units; recalculate if necessary.
    2. Eliminate outrageous answers, e.g., if there are two negative answers and two positive answers and the answer to the question can only be positive, cross out the two negative answers.
    3. Follow the SKIP advice above and come back.

- If there is a multiple-choice selection for "all of the above", and you have already found two of the answers, trust that you've done everything correctly, choose "all of the above", and move to the next problem. Again, the soon-to-be-engineer in you will tell you to check the final solution, but there's a great chance you did it correctly and you'll need the time!

- Consider using the answers from multiple-choice selections to back-calculate (or reverse engineer, as it were). This situation doesn't usually come up, but it does happen and it might allow you to eliminate a couple choices.

- Don't start any calculations until you have circled/underlined the actual problem statement (see below for a quick checklist for each problem on the exam).

- If you have time at the end, quickly check your circled solutions in the test booklet against the score sheet.

- Do not leave any answers blank. The exam proctors will inform the entire room at 15 minutes, at 5 minutes, and again at "pencils down". Use the last 15 minutes to make sure you have filled in all your answer bubbles.

PROBLEMS

PROBLEM 1

A vehicle's max speed (V_{max}) is 190 mph. The superelevation rate (e) for a highway horizontal curve is between 0.05 and 0.11. The side friction factor (f_s) is between is 0.09 and 0.16. The design speed (V_{mph}) for the curve is 55 mph. Find the maximum degree of curvature ($D°$) for the design of the curve.

Process for answering the problem (quickly write these down for each problem):

1. What type of problem is it? (5 seconds)

2. What is the problem asking for? (10 seconds)

3. Is there extraneous information? (10 seconds)

4. What references or equations are needed? (10 seconds)

5. Use the space on this sheet or scratch paper to find your solution.

6. Circle the right answer or write down the correct letter (do not fill in the answer sheet bubble yet).

7. If the answer does not appear, see passthecivilPE Exam Advice.

A) 7.67°
B) 3.61°
C) 5.12°
D) 8.49°

PROBLEM 2

A round column supporting an overpass near the south entrance of the Golden Gate Bridge in San Francisco is 6 ft in diameter. The column is scheduled to be wrapped with an advanced fiber and embedded epoxy resin composite material to increase the shear capacity and develop ductile performance during a seismic event. The specifications require the composite to be 15 ft high and wrapped in two layers. Required overlap of the wrap will account for an additional 5% of the total composite used. The fiber system selected is an E-glass fiberglass. The weight of the epoxy resin specified is 0.8 lb/ft^2 and costs $3/lb. What is the cost of the epoxy only for one column?

Process for answering the problem (quickly write these down for each problem):

1. What type of problem is it? (5 seconds)

2. What is the problem asking for? (10 seconds)

3. Is there extraneous information? (10 seconds)

4. What references or equations are needed? (10 seconds)

5. Use the space on this sheet or scratch paper to find your solution.

6. Circle the right answer or write down the correct letter (do not fill in the answer sheet bubble yet).

7. If the answer does not appear, see passthecivilPE Exam Advice.

A) $10,165.00
B) $9,925.80
C) $1,425.00
D) $5,314.00

© 2015 | passthecivilPE.com

PROBLEM 3

A reinforced concrete beam cross-section is shown below. The compressive strength of the concrete (f_c) is 2,800 psi. The yield strength of the tension steel (f_y) is 55 ksi. 3 - #8 reinforcing steel bars (*rebars*) will be used. The beam has 2 in. clear cover. The width (b) is 14 in. and the height (h) is 18 in. Find the area of concrete (A_c) required for a balanced condition.

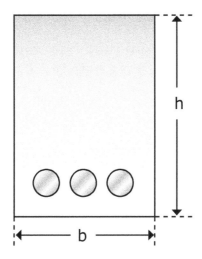

Process for answering the problem (quickly write these down for each problem):

1. What type of problem is it? (5 seconds)

2. What is the problem asking for? (10 seconds)

3. Is there extraneous information? (10 seconds)

4. What references or equations are needed? (10 seconds)

5. Use the space on this sheet or scratch paper to find your solution.

6. Circle the right answer or write down the correct letter
   (do not fill in the answer sheet bubble yet).

7. If the answer does not appear, see passthecivilPE Exam Advice.

A) 70.00 in²
B) 2.37 in²
C) 8.50 in²
D) 54.77 in²

PROBLEM 4

Which of the following is false concerning Project Planning? Choose the best answer.

A. The "float" for an activity in a project is equal to the difference between the late start and the early start for a particular activity

B. A "dummy" activity is a zero-duration activity used to demonstrate the logical dependence of an activity with respect to other project activities

C. More than one "critical path" may exist in a project network

D. The "critical path" is the shortest duration continuous path through a project network

Process for answering the problem (quickly write these down for each problem):

1. What type of problem is it? (5 seconds)

2. What is the problem asking for? (10 seconds)

3. Is there extraneous information? (10 seconds)

4. What references or equations are needed? (10 seconds)

5. Use the space on this sheet or scratch paper to find your solution.

6. Circle the right answer or write down the correct letter (do not fill in the answer sheet bubble yet).

7. If the answer does not appear, see passthecivilPE Exam Advice.

A) B
B) C
C) A
D) D

PROBLEM

Two closed cylindrical water tanks will be painted on the top and exterior surfaces. The bottom exterior surface and the interior of the tanks will not be painted. The tanks are 15 ft high and have a 25 ft diameter. There is a 3 ft x 3 ft square door used for all piping and access on the top surface of each water tank that will not be painted. The tanks require 2 coats of paint each. The paint covers 200 ft²/gallon and costs $35/gallon. The paint production rate is 675 ft²/hour and the job will use 2 workers, each paid $30/hour. The workers' pay should be rounded up to the nearest 1/2 hour at job completion. Neglect any other material costs (e.g., paint brushes, ladders). What is the cost for labor and material for this job?

Process for answering the problem (quickly write these down for each problem):

1. What type of problem is it? (5 seconds)

2. What is the problem asking for? (10 seconds)

3. Is there extraneous information? (10 seconds)

4. What references or equations are needed? (10 seconds)

5. Use the space on this sheet or scratch paper to find your solution.

6. Circle the right answer or write down the correct letter
(do not fill in the answer sheet bubble yet).

7. If the answer does not appear, see passthecivilPE Exam Advice.

A) $1,462.00
B) $2,925.50
C) $456.00
D) $8,641.00

PROBLEM 6

Which of the following is NOT a concrete admixture:

- A. Air entraining

- B. Water reducing

- C. Corrosion inhibiting

- D. Shrinkage reducing

- E. Organic growth promoting

- F. Retarding (setting time reduction)

- G. Accelerating (setting time accelerator)

Process for answering the problem (quickly write these down for each problem):

1. What type of problem is it? (5 seconds)

2. What is the problem asking for? (10 seconds)

3. Is there extraneous information? (10 seconds)

4. What references or equations are needed? (10 seconds)

5. Use the space on this sheet or scratch paper to find your solution.

6. Circle the right answer or write down the correct letter
    (do not fill in the answer sheet bubble yet).

7. If the answer does not appear, see passthecivilPE Exam Advice.

A) B and D
B) E only
C) F only
D) none of the above (all are concrete admixtures)

PROBLEM 7

Find the correct moment diagram for the loading shown below.

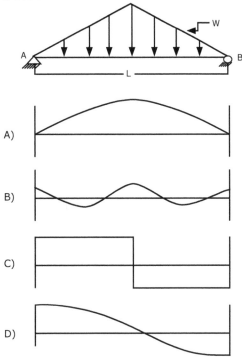

A)

B)

C)

D)

Process for answering the problem (quickly write these down for each problem):

1. What type of problem is it? (5 seconds)

2. What is the problem asking for? (10 seconds)

3. Is there extraneous information? (10 seconds)

4. What references or equations are needed? (10 seconds)

5. Use the space on this sheet or scratch paper to find your solution.

6. Circle the right answer or write down the correct letter
   (do not fill in the answer sheet bubble yet).

7. If the answer does not appear, see passthecivilPE Exam Advice.

A) C
B) A
C) D
D) B

PROBLEM

The moisture content for the fine and coarse aggregate for a concrete mix are 5.4% and 1.9%, respectively. The saturated surface dry (*SSD*) specific gravity (*SG*) for the fine and coarse aggregate are 2.69 and 2.63, respectively. The saturated surface dry (*SSD*) moisture content (ω_{SSD}) for the fine and coarse aggregate are 1.5% and 0.9%, respectively. The volume of a particular batch of concrete is 2.35 cf. Sulphates are not present where the concrete will be installed. The following table shows the weights of several crucial ingredients for the total concrete mix:

Weight of Water (lb)	Weight of Cement (lb)	Weight of Course Aggregate (lb)	Weight of Fine Aggregate (lb)
35.6	57.2	145.4	109.5

Find the water/cement ratio (w/c) for the concrete mix batch.

Process for answering the problem (quickly write these down for each problem):

1. What type of problem is it? (5 seconds)

2. What is the problem asking for? (10 seconds)

3. Is there extraneous information? (10 seconds)

4. What references or equations are needed? (10 seconds)

5. Use the space on this sheet or scratch paper to find your solution.

6. Circle the right answer or write down the correct letter (do not fill in the answer sheet bubble yet).

7. If the answer does not appear, see passthecivilPE Exam Advice.

A) 4.10 $\frac{gallons}{sack}$
B) 8.09 $\frac{gallons}{sack}$
C) 6.07 $\frac{gallons}{sack}$
D) 6.90 $\frac{gallons}{sack}$

PROBLEM 9

A 22 ft beam has a pinned connection at one end (*at A*) and a roller at the other end (*at B*), as shown below. The beam is subjected to a uniform load (W) of 0.6 kip/ft and a point load (*P*) of 5.5 kips that is 5 ft from the left end. The beam is 2 x 6 Douglas Fir. Find the location of maximum moment (M_{max}).

6 in.

2 in.

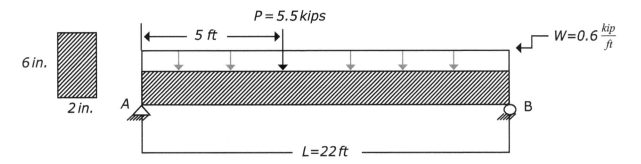

Process for answering the problem (quickly write these down for each problem):

1. What type of problem is it? (5 seconds)

2. What is the problem asking for? (10 seconds)

3. Is there extraneous information? (10 seconds)

4. What references or equations are needed? (10 seconds)

5. Use the space on this sheet or scratch paper to find your solution.

6. Circle the right answer or write down the correct letter (do not fill in the answer sheet bubble yet).

7. If the answer does not appear, see passthecivilPE Exam Advice.

A) 11.00 ft from A
B) 7.26 ft from A
C) 5.52 ft from A
D) 8.92 ft from A

PROBLEM 10

35 full-time employees (40 hrs/week and 2000 hrs/yr) and 20 half-time employees (20 hrs/week and 1000 hrs/yr) work for a concrete construction company. 4 half-time employees were treated with first aid for minor cuts and scrapes, 3 half-time employees had injuries resulting in "light duty", 3 full-time employees were medically treated, but had no time-off or "light duty", and 2 full-time employees were injured and took time-off. Find the OSHA Recordable Incident Rate (*IR*).

Process for answering the problem (quickly write these down for each problem):

1. What type of problem is it? (5 seconds)

2. What is the problem asking for? (10 seconds)

3. Is there extraneous information? (10 seconds)

4. What references or equations are needed? (10 seconds)

5. Use the space on this sheet or scratch paper to find your solution.

6. Circle the right answer or write down the correct letter
   (do not fill in the answer sheet bubble yet).

7. If the answer does not appear, see passthecivilPE Exam Advice.

A) 22.22%
B) 17.78%
C) 3.11%
D) 8.54%

PROBLEM 11

A university transportation engineering class conducted a traffic count of a major arterial road during evening rush-hour traffic. Find the peak hour factor (*PHF*) using the given vehicle volume and time interval.

Time Interval	Volume (Vehicles)	Time Interval	Volume (Vehicles)
4:30pm – 4:45pm	390	5:30pm – 5:45pm	575
4:45pm – 5:00pm	410	5:45pm – 6:00pm	560
5:00pm – 5:15pm	540	6:00pm – 6:15pm	420
5:15pm – 5:30pm	570	6:15pm – 6:30pm	400

Process for answering the problem (quickly write these down for each problem):

1. What type of problem is it? (5 seconds)

2. What is the problem asking for? (10 seconds)

3. Is there extraneous information? (10 seconds)

4. What references or equations are needed? (10 seconds)

5. Use the space on this sheet or scratch paper to find your solution.

6. Circle the right answer or write down the correct letter
   (do not fill in the answer sheet bubble yet).

7. If the answer does not appear, see passthecivilPE Exam Advice.

A) 2.394
B) 1.050
C) 0.980
D) 0.667

PROBLEM 12

A concrete "gravity" retaining wall with normal weight concrete ($\Upsilon_c=155$ lb/ft^3) holds soil backfill ($\Upsilon_s=120$ lb/ft^3), as shown below. The wall is 10 ft in height (H = 10 ft), 3 ft thick at the top, and 5 ft thick at the bottom, and 20 ft long. $\beta=0°$, $\alpha=0°$, $\varnothing=36°$, and the cohesion of the soil is $c=0°$. Ignore the exterior vertical wall friction, $\delta=0°$. However, assume the exterior friction under the base of the wall is $\delta_{base}=26°$. The water table is 2 ft from the top of the wall ($h = 2$ ft). Find the Factor of Safety for sliding (FOS) for the retaining wall, using the Rankine Theory for active earth pressure.

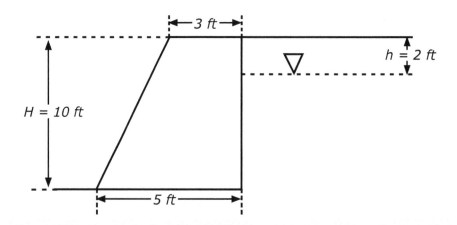

Process for answering the problem (quickly write these down for each problem):

1. What type of problem is it? (5 seconds)

2. What is the problem asking for? (10 seconds)

3. Is there extraneous information? (10 seconds)

4. What references or equations are needed? (10 seconds)

5. Use the space on this sheet or scratch paper to find your solution.

6. Circle the right answer or write down the correct letter
   (do not fill in the answer sheet bubble yet).

7. If the answer does not appear, see passthecivilPE Exam Advice.

A) 4.00
B) 1.00
C) 1.70
D) 2.65

PROBLEM 13

The lowest point of an overpass is required to be 24 ft above the high point of a crest curve, as shown. The distance between the PT and the PI is 400 ft. The elevation of the PI is 1100 ft. Find the elevation of the high point of the curve (Y_{HP}).

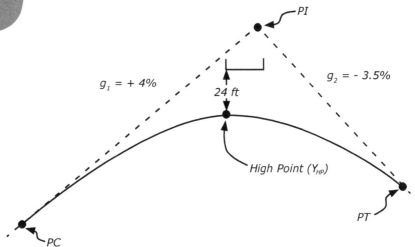

Process for answering the problem (quickly write these down for each problem):

1. What type of problem is it? (5 seconds)

2. What is the problem asking for? (10 seconds)

3. Is there extraneous information? (10 seconds)

4. What references or equations are needed? (10 seconds)

5. Use the space on this sheet or scratch paper to find your solution.

6. Circle the right answer or write down the correct letter (do not fill in the answer sheet bubble yet).

7. If the answer does not appear, see passthecivilPE Exam Advice.

A) 1100.00 ft
B) 950.74 ft
C) 1092.53 ft
D) 678.69 ft

PROBLEM 14

A concrete retaining wall with normal weight concrete ($\Upsilon_c = 155$ lb/ft³) holds soil backfill ($\Upsilon_s = 120$ lb/ft³), as shown below. The wall is 12 ft in height, 2 ft thick at the top, 3 ft thick at the bottom, and 20 ft long. The friction angle between the soil and the concrete (δ) is 21°. The angle of the soil backfill to the horizontal (β) is 1V:2H. The angle of the surface of the retaining wall to the vertical (α) is 15°. The angle of internal friction ($\varnothing$) is 31°. The cohesion of the soil (c) is 0°. Ignore friction under the wall. Draw the angles into the diagram below and find the active earth pressure coefficient.

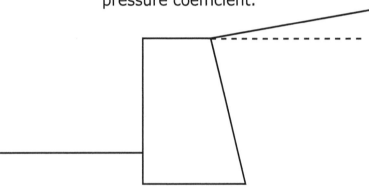

Process for answering the problem (quickly write these down for each problem):

1. What type of problem is it? (5 seconds)

2. What is the problem asking for? (10 seconds)

3. Is there extraneous information? (10 seconds)

4. What references or equations are needed? (10 seconds)

5. Use the space on this sheet or scratch paper to find your solution.

6. Circle the right answer or write down the correct letter (do not fill in the answer sheet bubble yet).

7. If the answer does not appear, see passthecivilPE Exam Advice.

A) 0.346
B) 0.767
C) - 0.767
D) - 0.902

PROBLEM 15

The lowest point of an overpass is required to be a minimum of 18 ft above point P on a vertical sag curve as shown. Find the minimum elevation of the overpass if station of the Point of Tangency (*PT*) is 17+00 and the station at a point on the curve (*P*) is 8+02. The distance between PC and PI is 525 ft. The elevation of PC is 255 ft, The elevation of PT is 400 ft.

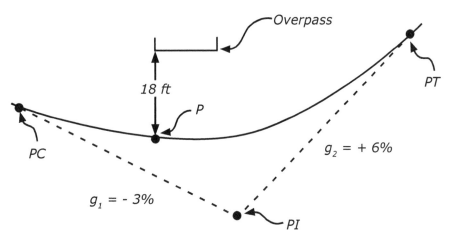

Process for answering the problem (quickly write these down for each problem):

1. What type of problem is it? (5 seconds)

2. What is the problem asking for? (10 seconds)

3. Is there extraneous information? (10 seconds)

4. What references or equations are needed? (10 seconds)

5. Use the space on this sheet or scratch paper to find your solution.

6. Circle the right answer or write down the correct letter (do not fill in the answer sheet bubble yet).

7. If the answer does not appear, see passthecivilPE Exam Advice.

A) 269.43 ft
B) 365.20 ft
C) 150 ft
D) 221.65 ft

PROBLEM 16

The activity-on-arrow for a bridge project in Yellowstone National Park is shown. Find the early start (*ES*) and early finish (*EF*) for activity K.

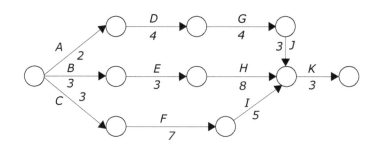

Activity	Duration
A	2
B	3
C	3
D	4

Activity	Duration
E	3
F	7
G	4
H	8

Activity	Duration
I	5
J	3
K	3

Process for answering the problem (quickly write these down for each problem):

1. What type of problem is it? (5 seconds)

2. What is the problem asking for? (10 seconds)

3. Is there extraneous information? (10 seconds)

4. What references or equations are needed? (10 seconds)

5. Use the space on this sheet or scratch paper to find your solution.

6. Circle the right answer or write down the correct letter (do not fill in the answer sheet bubble yet).

7. If the answer does not appear, see passthecivilPE Exam Advice.

A) ES = 17, EF = 18
B) ES = 18, EF = 15
C) ES = 15, EF = 18
D) ES = 15, EF = 20

PROBLEM 17

A vertical crest curve has a +2% and a -6% grade as shown below. The distance between the Point of Curvature (*PC*) and the Point of Intersection (*PI*) is 1,125 ft. The station of the Point of Tangency (*PT*) is 42+25. The station at a Point on the curve (*P*) is 28+75. The elevation at PC (Y_{PC}) is 100 ft. Find the slope at Point P (S_P).

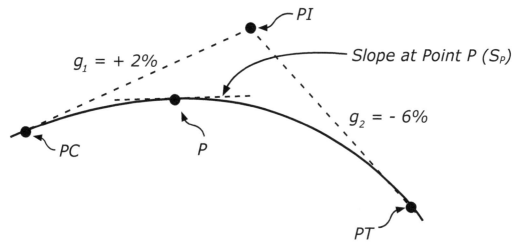

Process for answering the problem (quickly write these down for each problem):

1. What type of problem is it? (5 seconds)

2. What is the problem asking for? (10 seconds)

3. Is there extraneous information? (10 seconds)

4. What references or equations are needed? (10 seconds)

5. Use the space on this sheet or scratch paper to find your solution.

6. Circle the right answer or write down the correct letter
(do not fill in the answer sheet bubble yet).

7. If the answer does not appear, see passthecivilPE Exam Advice.

A) -1.2 %
B) 2.1 %
C) 1.1 %
D) 0.9 %

PROBLEM 18

Modular steel frame formwork is used for a retaining wall that is 11.5 ft in height. The temperature outside is 68°F. The temperature of the concrete is 75°F. The cement contains 44% fly ash and the concrete has density of 155 lb/ft³. The concrete will be placed at 7.5 ft/hr. Find the lateral pressure of the concrete on the formwork using ACI 347.

Process for answering the problem (quickly write these down for each problem):

1. What type of problem is it? (5 seconds)

2. What is the problem asking for? (10 seconds)

3. Is there extraneous information? (10 seconds)

4. What references or equations are needed? (10 seconds)

5. Use the space on this sheet or scratch paper to find your solution.

6. Circle the right answer or write down the correct letter
   (do not fill in the answer sheet bubble yet).

7. If the answer does not appear, see passthecivilPE Exam Advice.

A) 1521 psf
B) 2511 psf
C) 1511 psf
D) 3511 psf

PROBLEM 19

Which of the following is not one of the primary intentions and purpose of the Occupational Safety and Health Administration (OSHA) regulations? Choose the best answer.

A) Any worker (with an employer regulated under the jurisdiction of OSHA) may file a complaint to have OSHA inspect their workplace if they believe that their employer is not following OSHA standards, or there are serious hazards in their workplace.

B) OSHA only applies to construction workers and work related to construction.

C) The Occupational Safety and Health Act of 1970 (OSH Act), which led to the creation of OSHA, requires employers to provide their employees with working conditions that are free of known dangers.

D) OSHA regulates construction worker safety separately from its general industry, maritime, and agricultural requirements, and provides construction industry regulations which can be found in the standard publication titled 29 CFR 1926.

Process for answering the problem (quickly write these down for each problem):

1. What type of problem is it? (5 seconds)

2. What is the problem asking for? (10 seconds)

3. Is there extraneous information? (10 seconds)

4. What references or equations are needed? (10 seconds)

5. Use the space on this sheet or scratch paper to find your solution.

6. Circle the right answer or write down the correct letter (do not fill in the answer sheet bubble yet).

7. If the answer does not appear, see passthecivilPE Exam Advice.

A) D
B) A
C) B
D) C

PROBLEM 20

A temporary structure has been built on a construction site in Frederick, Maryland. The structure will be in place for approximately 1.5 years. The design wind speed used for permanent structures adjacent to the temporary structure is 85 mph. Find the lateral wind pressure used to design the temporary structure using ASCE 37.

Process for answering the problem (quickly write these down for each problem):

1. What type of problem is it? (5 seconds)

2. What is the problem asking for? (10 seconds)

3. Is there extraneous information? (10 seconds)

4. What references or equations are needed? (10 seconds)

5. Use the space on this sheet or scratch paper to find your solution.

6. Circle the right answer or write down the correct letter (do not fill in the answer sheet bubble yet).

7. If the answer does not appear, see passthecivilPE Exam Advice.

A) 6.24 psf
B) 14.53 psf
C) 24.67 psf
D) 13.36 psf

PROBLEM 21

Which of the following is false regarding open channel flow? Choose the best answer.

A) In general, the most efficient open channel cross section will have a maximized hydraulic radius, R, and a minimized wetted perimeter, P.

B) The usual minimum permissible velocity of an open channel, including closed-conduit pipe networks flowing under the influence of gravity only, is the lowest velocity that prevents sediment deposit.

C) The liquid sheet Nappe of a rectangular weir decreases in width as it falls if the weir opening width is less than the channel width.

D) Laminar flow occurs at high Reynolds numbers, and turbulent flow occurs at low Reynolds numbers.

Process for answering the problem (quickly write these down for each problem):

1. What type of problem is it? (5 seconds)

2. What is the problem asking for? (10 seconds)

3. Is there extraneous information? (10 seconds)

4. What references or equations are needed? (10 seconds)

5. Use the space on this sheet or scratch paper to find your solution.

6. Circle the right answer or write down the correct letter
    (do not fill in the answer sheet bubble yet).

7. If the answer does not appear, see passthecivilPE Exam Advice.

A) A
B) C
C) D
D) B

PROBLEM 22

A new water tower concrete foundation slab specification requires modified proctor testing to establish the soil compaction parameters. Five tests were performed with the following results at the laboratory.

Test	Weight	Moisture Content (ω)
1	5.72 lb	8.3%
2	5.96 lb	8.9%
3	6.15 lb	9.3%
4	6.13 lb	9.5%
5	5.89 lb	10.3%

At the construction site, 0.02 ft³ of compacted soil is tested and has a wet weight of 2.4 lb and dry weight of 2.1 lb. Find the percentage of compaction for the in-situ soil.

Process for answering the problem (quickly write these down for each problem):

1. What type of problem is it? (5 seconds)

2. What is the problem asking for? (10 seconds)

3. Is there extraneous information? (10 seconds)

4. What references or equations are needed? (10 seconds)

5. Use the space on this sheet or scratch paper to find your solution.

6. Circle the right answer or write down the correct letter (do not fill in the answer sheet bubble yet).

7. If the answer does not appear, see passthecivilPE Exam Advice.

A) 100%
B) 61.8%
C) 85.3%
D) 78.9%

23

PROBLEM

Due to a setback issue near the city of Bakersfield, California, an open channel section has the dimensions shown below. The slope of the concrete channel is 0.0004. Find the hydraulic radius (R) of the channel section.

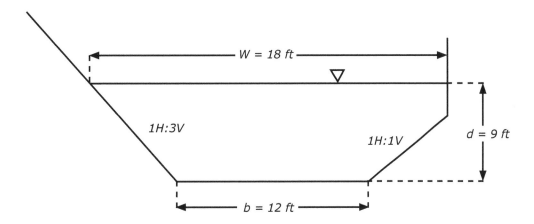

Process for answering the problem (quickly write these down for each problem):

1. What type of problem is it? (5 seconds)

2. What is the problem asking for? (10 seconds)

3. Is there extraneous information? (10 seconds)

4. What references or equations are needed? (10 seconds)

5. Use the space on this sheet or scratch paper to find your solution.

6. Circle the right answer or write down the correct letter (do not fill in the answer sheet bubble yet).

7. If the answer does not appear, see passthecivilPE Exam Advice.

A) 6.54 ft
B) 5.45 ft
C) 4.54 ft
D) 7.45 ft

PROBLEM 24

Choose the answer below that is false:

A) The difference between liquid limit (*LL*) and the plastic limit (*PL*) is defined as the plasticity index (*PI*).

B) When the weight of water equals the weight of the dry soil (i.e., $\omega = 100\%$) in a soil sample the liquid limit (*LL*) is 100.

C) The plastic limit of a soil sample is attained when a soil sample that is rolled in to an 1/8 in. diameter thread begins to crumble.

D) Atterberg limit tests can be applied to other construction materials such as cement mixtures and certain asphaltic materials.

Process for answering the problem (quickly write these down for each problem):

1. What type of problem is it? (5 seconds)

2. What is the problem asking for? (10 seconds)

3. Is there extraneous information? (10 seconds)

4. What references or equations are needed? (10 seconds)

5. Use the space on this sheet or scratch paper to find your solution.

6. Circle the right answer or write down the correct letter (do not fill in the answer sheet bubble yet).

7. If the answer does not appear, see passthecivilPE Exam Advice.

A) A and C
B) D only
C) B and D
D) A only

PROBLEM 25

A pump is used to dewater the soil adjacent to an elevator pit. The top of the concrete slab at the base of the elevator pit is 4 ft under the water table. The concrete slab is 3 ft thick. The walls of the elevator pit are concrete with blindside waterproofing and wood lagging with steel soldier piles. The water is pumped to a basin 20 ft above the top of the concrete slab. The specifications require the water table to be at least 2 ft below the bottom of slab. The pipe has a diameter of 6 in., length of 200 ft, and a Darcy Friction Factor of 0.04. The pump efficiency is 85% and the flow rate is 100 gallons per minute. Find the horsepower required for the pump.

Process for answering the problem (quickly write these down for each problem):

1. What type of problem is it? (5 seconds)

2. What is the problem asking for? (10 seconds)

3. Is there extraneous information? (10 seconds)

4. What references or equations are needed? (10 seconds)

5. Use the space on this sheet or scratch paper to find your solution.

6. Circle the right answer or write down the correct letter
   (do not fill in the answer sheet bubble yet).

7. If the answer does not appear, see passthecivilPE Exam Advice.

A) 1.954 *hp*
B) 3.854 *hp*
C) 0.075 *hp*
D) 0.754 *hp*

PROBLEM 26

The specification for an airport access road in Sacramento, California, requires standard proctor compaction. The wet soil sample is 0.29 lbs and oven dried soil sample is 0.25 lbs. The wet soil sample and mold weighs 13.2 lbs and the empty mold weighs 9.5 lbs. Find the in-situ dry unit weight (Υ_d) of the soil.

Process for answering the problem (quickly write these down for each problem):

1. What type of problem is it? (5 seconds)

2. What is the problem asking for? (10 seconds)

3. Is there extraneous information? (10 seconds)

4. What references or equations are needed? (10 seconds)

5. Use the space on this sheet or scratch paper to find your solution.

6. Circle the right answer or write down the correct letter (do not fill in the answer sheet bubble yet).

7. If the answer does not appear, see passthecivilPE Exam Advice.

A) $98.07 \frac{lb}{ft^3}$

B) $95.69 \frac{lb}{ft^3}$

C) $73.62 \frac{lb}{ft^3}$

D) $70.50 \frac{lb}{ft^3}$

PROBLEM 27

Which of the following is false regarding slope stability at a construction site? Choose the best answer.

A) In general, unshored sides of soil should not be steeper than their in-situ angle of repose.

B) The most probable type of slip-plane for a trench in soft mud with walls sloped at 55° is "Toe Circle".

C) Slope failures typically occur in the shape of a sliding rectangle, where the depth of the sliding soil is approximately the same thickness at the top, middle, and toe of the slide.

D) In general, if the forces available within the soil to resist movement are greater than the forces driving movement, the slope is considered stable.

Process for answering the problem (quickly write these down for each problem):

1. What type of problem is it? (5 seconds)

2. What is the problem asking for? (10 seconds)

3. Is there extraneous information? (10 seconds)

4. What references or equations are needed? (10 seconds)

5. Use the space on this sheet or scratch paper to find your solution.

6. Circle the right answer or write down the correct letter
   (do not fill in the answer sheet bubble yet).

7. If the answer does not appear, see passthecivilPE Exam Advice.

A) A
B) B
C) C
D) D

PROBLEM 28

The mass diagram is shown below for a section of a new highway project in Pittsburgh, Pennsylvania. Choose the best answer below for which is true, given the mass diagram shown.

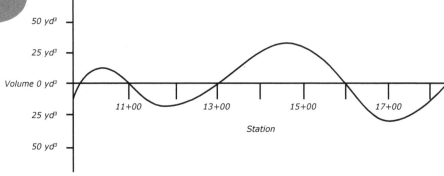

A) The section between Station 13+00 and 16+00 represents a fill operation.

B) The total grading operation is balanced.

C) The section between Station 11+00 and 16+00 represents the "free-haul distance".

D) The section at Station 17+00 represents a transition between fill and cut.

Process for answering the problem (quickly write these down for each problem):

1. What type of problem is it? (5 seconds)

2. What is the problem asking for? (10 seconds)

3. Is there extraneous information? (10 seconds)

4. What references or equations are needed? (10 seconds)

5. Use the space on this sheet or scratch paper to find your solution.

6. Circle the right answer or write down the correct letter (do not fill in the answer sheet bubble yet).

7. If the answer does not appear, see passthecivilPE Exam Advice.

A) A
B) B and D
C) C
D) A and C

PROBLEM 29

Which of the following is false regarding temporary erosion control at a construction site? Choose the best answer.

A) In general, sand bags should be placed where they can divert and slow water and sediment flow to accumulate into predetermined deposit locations.

B) Silt fences, straw bale barriers, and sand bag barriers are all considered temporary erosion control.

C) Temporary erosion control is mandatory only at construction projects that require below-grade and slab-on-grade concrete operations.

D) The desired objective when developing a temporary erosion control plan is to keep soil at its original location.

Process for answering the problem (quickly write these down for each problem):

1. What type of problem is it? (5 seconds)

2. What is the problem asking for? (10 seconds)

3. Is there extraneous information? (10 seconds)

4. What references or equations are needed? (10 seconds)

5. Use the space on this sheet or scratch paper to find your solution.

6. Circle the right answer or write down the correct letter (do not fill in the answer sheet bubble yet).

7. If the answer does not appear, see passthecivilPE Exam Advice.

A) D
B) A
C) C
D) B

PROBLEM 30

A saturated soil sample has a dry unit weight (Υ_d) of 165 lb/ft³ and a water content (ω) of 12.5%. The sample is borrow soil to be used for a cantilever wall foundation. Find the void ratio (e) of the soil sample.

Process for answering the problem (quickly write these down for each problem):

1. What type of problem is it? (5 seconds)

2. What is the problem asking for? (10 seconds)

3. Is there extraneous information? (10 seconds)

4. What references or equations are needed? (10 seconds)

5. Use the space on this sheet or scratch paper to find your solution.

6. Circle the right answer or write down the correct letter (do not fill in the answer sheet bubble yet).

7. If the answer does not appear, see passthecivilPE Exam Advice.

A) 1.000
B) 0.331
C) 0.009
D) 0.978

PROBLEM 31

A 12 ft wide rectangular channel produces a hydraulic jump from supercritical flow to subcritical flow. The depth at the point where the hydraulic jump occurs is 1.5 ft. The supercritical flow rate is 185 ft³/sec. Find the velocity of the water after the hydraulic jump.

Process for answering the problem (quickly write these down for each problem):

1. What type of problem is it? (5 seconds)

2. What is the problem asking for? (10 seconds)

3. Is there extraneous information? (10 seconds)

4. What references or equations are needed? (10 seconds)

5. Use the space on this sheet or scratch paper to find your solution.

6. Circle the right answer or write down the correct letter
   (do not fill in the answer sheet bubble yet).

7. If the answer does not appear, see passthecivilPE Exam Advice.

A) 9.35 ft/s
B) 1.75 ft/s
C) 0.27 ft/s
D) 6.22 ft/s

PROBLEM 32

A 15 ft hollow meter beam weighs 1,575 lbs, has a pinned connection at one end, a roller at the other end, and the cross-section shown below. The statical moment of area about the neutral axis (Q) is = 76.0 in³. Find the moment of inertia (I).

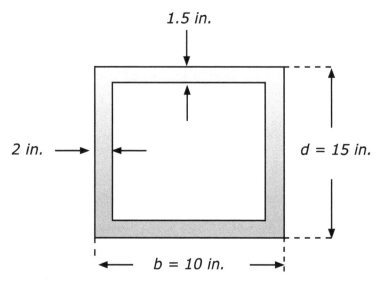

Process for answering the problem (quickly write these down for each problem):

1. What type of problem is it? (5 seconds)

2. What is the problem asking for? (10 seconds)

3. Is there extraneous information? (10 seconds)

4. What references or equations are needed? (10 seconds)

5. Use the space on this sheet or scratch paper to find your solution.

6. Circle the right answer or write down the correct letter (do not fill in the answer sheet bubble yet).

7. If the answer does not appear, see passthecivilPE Exam Advice.

A) 1242.0 *in⁴*
B) 2311.5 *in⁴*
C) 1948.5 *in⁴*
D) 1799.0 *in⁴*

PROBLEM 33

Which of the following is false regarding hydrology for stormwater runoff? Choose the best answer.

A) The "Horton Method" and "Green–Ampt Method" are methods for estimating ground water infiltration rates.

B) The Runoff Curve Number (or Curve Number, CN) is used in determining the approximate amount of direct runoff or ground water infiltration from a rainfall event.

C) A detention pond and a retention pond are the same type of runoff basin.

D) The Time of Concentration of a watershed is the time needed for water to flow from the most remote point in a watershed to the watershed outlet.

Process for answering the problem (quickly write these down for each problem):

1. What type of problem is it? (5 seconds)

2. What is the problem asking for? (10 seconds)

3. Is there extraneous information? (10 seconds)

4. What references or equations are needed? (10 seconds)

5. Use the space on this sheet or scratch paper to find your solution.

6. Circle the right answer or write down the correct letter
    (do not fill in the answer sheet bubble yet).

7. If the answer does not appear, see passthecivilPE Exam Advice.

A) A
B) B
C) C
D) D

PROBLEM 34

An 18 ft Douglas Fir timber beam has a pinned connection at one end (*at A*) and a roller at the other end (*at B*), as shown below. The beam is subjected to a uniform load (*W*) of 150 lb/ft and a point load (*P*) of 1.2 kips at midspan. The beam is 4 in. wide and 12 in. deep. Find the maximum flexural stress (f_b), in ksi.

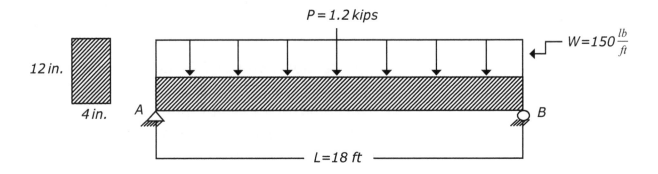

Process for answering the problem (quickly write these down for each problem):

1. What type of problem is it? (5 seconds)

2. What is the problem asking for? (10 seconds)

3. Is there extraneous information? (10 seconds)

4. What references or equations are needed? (10 seconds)

5. Use the space on this sheet or scratch paper to find your solution.

6. Circle the right answer or write down the correct letter (do not fill in the answer sheet bubble yet).

7. If the answer does not appear, see passthecivilPE Exam Advice.

A) 4.67 ksi
B) 1.44 ksi
C) 2.50 ksi
D) 0.56 ksi

PROBLEM 35

A welded and seamless steel pipe with 33 in. circumference provides water from a large water tank on a hill above several hotels near Yosemite National Park in California. The water tank is 700 ft above the subdivision. The pipe system is approximately 3/4 mile away. The flow inside the pipe is determined to be 3.5 ft^3/sec. Find the head loss due to friction (h_f) of the pipe between the tank and the subdivision, in feet.

Process for answering the problem (quickly write these down for each problem):

1. What type of problem is it? (5 seconds)

2. What is the problem asking for? (10 seconds)

3. Is there extraneous information? (10 seconds)

4. What references or equations are needed? (10 seconds)

5. Use the space on this sheet or scratch paper to find your solution.

6. Circle the right answer or write down the correct letter
   (do not fill in the answer sheet bubble yet).

7. If the answer does not appear, see passthecivilPE Exam Advice.

A) 80.6 ft
B) 72.5 ft
C) 100.2 ft
D) 250.0 ft

PROBLEM 36

A truss system has a pinned connection on one side (*at Point A*) and a roller on the other side (*at Point H*), as shown below. 2 kips acts downward at Point C and 5 kips acts downward at Point E and Point F. Neglect dead load of the framing members. Find the force in member BE.

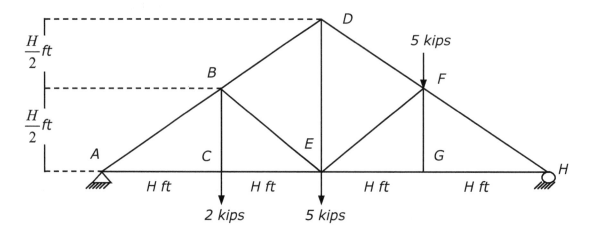

Process for answering the problem (quickly write these down for each problem):

1. What type of problem is it? (5 seconds)

2. What is the problem asking for? (10 seconds)

3. Is there extraneous information? (10 seconds)

4. What references or equations are needed? (10 seconds)

5. Use the space on this sheet or scratch paper to find your solution.

6. Circle the right answer or write down the correct letter (do not fill in the answer sheet bubble yet).

7. If the answer does not appear, see passthecivilPE Exam Advice.

A) 1.54 kips (tension)
B) 3.52 kips (tension)
C) 2.24 kips (compression)
D) 1.05 kips (compression)

PROBLEM 37

Which of the following is false regarding soil stress at a construction site? Choose the best answer.

A) Effective stress in soils is the difference between the total stress and the pore water pressure.

B) Above the water table, total and effective stresses are equal.

C) Below the water table, the pore pressure is a product of the specific weight of water and the height of the water above the point measured.

D) In dry soil, particles at a point underground experience a total overhead stress (depth underground multiplied by the specific weight of the soil) that is less than the stress at the same point in saturated soil.

Process for answering the problem (quickly write these down for each problem):

1. What type of problem is it? (5 seconds)

2. What is the problem asking for? (10 seconds)

3. Is there extraneous information? (10 seconds)

4. What references or equations are needed? (10 seconds)

5. Use the space on this sheet or scratch paper to find your solution.

6. Circle the right answer or write down the correct letter (do not fill in the answer sheet bubble yet).

7. If the answer does not appear, see passthecivilPE Exam Advice.

A) A
B) B
C) C
D) D

PROBLEM 38

Timber shoring and formwork support a 6 in. concrete slab (*150 pcf*), as shown below. The concrete will be cast-in-place over 3/4 in. plywood sheathing (*2.3 psf*). The 6 x 6 shores are spaced @ 48 in. o.c. The 4 x 6 stringers are spaced @ 5 ft o.c. The 2 x 4 joists (*1.5 in. x 3.5 in. actual*) are spaced at 16 in. o.c. The live load will be 20 psf. Assume the dead load is only the concrete and the plywood. Some lateral cross-bracing is installed for the shores. Metal bracket connections mechanically attach the shores to sill plates. Find the maximum bending stress (σ_b) on the joists.

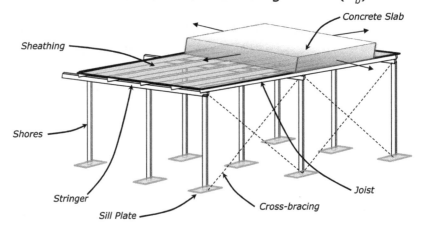

Sheathing · Concrete Slab · Shores · Stringer · Sill Plate · Cross-bracing · Joist

Process for answering the problem (quickly write these down for each problem):

1. What type of problem is it? (5 seconds)

2. What is the problem asking for? (10 seconds)

3. Is there extraneous information? (10 seconds)

4. What references or equations are needed? (10 seconds)

5. Use the space on this sheet or scratch paper to find your solution.

6. Circle the right answer or write down the correct letter
   (do not fill in the answer sheet bubble yet).

7. If the answer does not appear, see passthecivilPE Exam Advice.

A) 1.6 ksi
B) 2.3 ksi
C) 3.2 ksi
D) 5.0 ksi

PROBLEM 39

A network of pipes is configured in parallel under a playground in San Diego, California, as shown. Two of the pipes are schedule–40 steel and one of the pipes is bituminous-lined cast iron. The nominal sizes of the pipes are 3 in., 4 in., and 6 in., with Hazen–Williams loss coefficients of 100, 100, and 140, as shown. The 3 in. pipe is 150 ft long. The 4 in. pipe is 90 ft long. The 6 in. pipe is 225 ft long. Minor losses are insignificant. Water enters the pipe system at A at 4.2 ft³/sec. Find the total friction loss between junctions A and B.

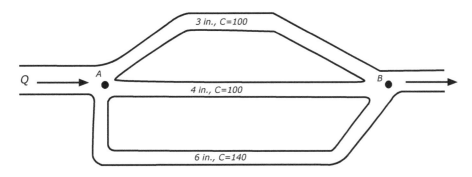

Process for answering the problem (quickly write these down for each problem):

1. What type of problem is it? (5 seconds)

2. What is the problem asking for? (10 seconds)

3. Is there extraneous information? (10 seconds)

4. What references or equations are needed? (10 seconds)

5. Use the space on this sheet or scratch paper to find your solution.

6. Circle the right answer or write down the correct letter (do not fill in the answer sheet bubble yet).

7. If the answer does not appear, see passthecivilPE Exam Advice.

A) -4.34 ft
B) 20.64 ft
C) 35.43 ft
D) 2.38 ft

PROBLEM 40

A 2,700 lb steam hammer is dropped from a height of 3.7 ft into soil with unit weight (Y_{soil}) of 110 lb/ft³. The water content (ω) is 8% and the specific gravity of the solids (SG_{solids}) is 2.67. The pile has driven 1.2 ft in the last 10 blows. Find the allowable capacity of the pile (Q_a) using the Engineering News (ENR) Formula if the driven weight is 2,995 lbs.

Process for answering the problem (quickly write these down for each problem):

1. What type of problem is it? (5 seconds)

2. What is the problem asking for? (10 seconds)

3. Is there extraneous information? (10 seconds)

4. What references or equations are needed? (10 seconds)

5. Use the space on this sheet or scratch paper to find your solution.

6. Circle the right answer or write down the correct letter
    (do not fill in the answer sheet bubble yet).

7. If the answer does not appear, see passthecivilPE Exam Advice.

A) 2,350 lbs
B) 35.5 kips
C) 12.9 kips
D) 420 lbs

© 2015 | passthecivilPE.com

SOLUTIONS

SOLUTION 1

A vehicle's max speed (V_{max}) is 190 mph. The superelevation rate (e) for a highway horizontal curve is between 0.05 and 0.11. The side friction factor (f_s) is between is 0.09 and 0.16. The design speed (V_{mph}) for the curve is 55 mph. Find the maximum degree of curvature ($D°$) for the design of the curve.

1. What type of problem is it?

    ■ Transportation: Geometrics

2. What is the problem asking for?

    ■ The question is asking for the Maximum Degree of Curvature ($D°$)

3. Is there extraneous information?

    ■ The max speed is not needed

4. What references or equations are needed?

    ■ $(D°) = \dfrac{5729.6}{R_{ft}}$, R is the Radius of Curvature, in feet

    ■ $R_{ft} = \dfrac{(V_{mph})^2}{15 \ (e + f_s)}$

5. Solution:

    STEP 1 $R_{ft} = \dfrac{(55_{mph})^2}{15 \ (0.11 + 0.16)} = 746.91 \ ft$

    STEP 2 $D° = \dfrac{5729.6}{R_{ft}} = 7.67°$

┤ **"THINGS TO REMEMBER"** ├
Note: by the equation, maximizing $e + f_s$, R is smaller. When R is smaller, $D°$ is maximized.

Correct Answer: (A)

SOLUTION 2

A round column supporting an overpass near the south entrance of the Golden Gate Bridge in San Francisco is 6 ft in diameter. The column is scheduled to be wrapped with an advanced fiber and embedded epoxy resin composite material to increase the shear capacity and develop ductile performance during a seismic event. The specifications require the composite to be 15 ft high and wrapped in two layers. Required overlap of the wrap will account for an additional 5% of the total composite used. The fiber system selected is an E-glass fiberglass. The weight of the epoxy resin specified is 0.8 lb/ft^2 and costs $3/lb. What is the cost of the epoxy only for one column?

1. What type of problem is it?

    ■ Construction: Project Planning and Cost Estimating

2. What is the problem asking for?

    ■ The final cost for the epoxy material for one column

3. Is there extraneous information?

    ■ The location and structure type (overpass) is not needed

4. What references or equations are needed?

    ■ Basic geometric equations

    ■ Reference Project Planning, Quantity Take-Off Methods and Cost Estimating

---

"QUICK TIPS"

When dealing with cost estimating and quantity take-off methods, be sure to utilize dimensional analysis to make sure units are accurate

---

5. Solution:

STEP 1 Calculate square footage:

Surface area of a cylinder without top and bottom is

2(π)rh = 2 (π)(3ft)(15ft) = 282.74 ft²

STEP 2 Calculate area for 2 layers:

(2)(282.74 ft²) = 565.48 ft²

STEP 3 Include overlap area:

(1.05)(565.48 ft²) = 593.75 ft²

STEP 4 Calculate weight:

(0.8 lb/ft²)(593.75 ft²) = 475 lb

STEP 5 Calculate epoxy cost:

($3/lb)(475 lb) = $1,425

Correct Answer: (C)

SOLUTION 3

A reinforced concrete beam cross-section is shown below. The compressive strength of the concrete (f_c) is 2,800 psi. The yield strength of the tension steel (f_y) is 55 ksi. 3 - #8 reinforcing steel bars (*rebars*) will be used. The beam has 2 in. clear cover. The width (*b*) is 14 in. and the height (*h*) is 18 in. Find the area of concrete (A_c) required for a balanced condition.

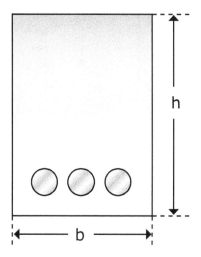

1. What type of problem is it?

   ■ Structural: Structural Mechanics

2. What is the problem asking for?

   ■ The area of concrete that is required for a balanced condition

3. Is there extraneous information?

   ■ The concrete clear cover is not needed to answer the question

4. What references or equations are needed?

   ■ Concrete beam loading

   ■ Concrete reinforcement

   ■ ACI 318 (Appendix E)

5. Solution:

STEP 1 Draw diagram to depict the balanced condition of the reinforcement steel and the compression area of the concrete, with strain, stress and forces:

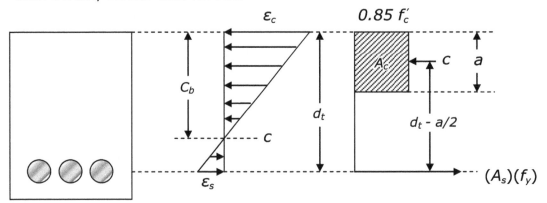

STEP 2 Find the equation for a balanced condition:

$$(0.85f_c')(A_c)=(A_s)(f_y)$$

STEP 3 Solve for the area of concrete in compression, using the equation for a balanced condition:

$$\therefore \ A_c \ = \frac{(A_s)(f_y)}{(0.85\,f_c')}$$

STEP 4 Using a reference to find typical rebar area, find the cross-section area (A_s) for #8 rebar:

A_s for #8 Bar = 0.79in²

$\therefore$ 3#8 rebars = (3)(0.79in²) = 2.37 in²

STEP 5 Find the area of concrete in compression:

$$A_c \ = \frac{\left(2.37\ in^2\right)\left(55,000\ \dfrac{lb}{in^2}\right)}{(0.85)\left(2,800\ \dfrac{lb}{in^2}\right)} = 54.77\ in^2$$ **Correct Answer: (D)**

Which of the following is false concerning Project Planning? Choose the best answer.

A. The "float" for an activity in a project is equal to the difference between the late start and the early start for a particular activity

B. A "dummy" activity is a zero-duration activity used to demonstrate the logical dependence of an activity with respect to other project activities

C. More than one "critical path" may exist in a project network

D. The "critical path" is the shortest duration continuous path through a project network

1. What type of problem is it?

   ■ Construction: Project Planning

2. What is the problem asking for?

   ■ The false choice regarding Project Planning

3. Is there extraneous information?

   ■ No

4. What references or equations are needed?

   ■ Reference project scheduling for construction projects

5. Solution:

"QUICK TIPS"

If there is only one false answer, then all the other answers are true, and may be useful on the exam.

The "critical path" is the longest duration continuous path of activities through a project, has the lowest "total float" for the project, and determines the project completion date.

Correct Answer: (D)

SOLUTION 5

Two closed cylindrical water tanks will be painted on the top and exterior surfaces. The bottom exterior surface and the interior of the tanks will not be painted. The tanks are 15 ft high and have a 25 ft diameter. There is a 3 ft x 3 ft square door used for all piping and access on the top surface of each water tank that will not be painted. The tanks require 2 coats of paint each. The paint covers 200 ft²/gallon and costs $35/gallon. The paint production rate is 675 ft²/hour and the job will use 2 workers, each paid $30/hour. The workers' pay should be rounded up to the nearest 1/2 hour at job completion. Neglect any other material costs (e.g., paint brushes, ladders). What is the cost for labor and material for this job?

1. What type of problem is it?

   ■ Construction: Project Planning

2. What is the problem asking for?

   ■ The cost for labor and material for the construction job

3. Is there extraneous information?

   ■ No

4. What references or equations are needed?

   ■ Dimensional analysis

   ■ Quantity take-off methods and cost estimating

5. Solution:

   STEP 1 Calculate square footage of 1 tank:

   Surface area of a cylinder with a top and without bottom is

   $$2(\pi)rh + (\pi)r^2 = 2(\pi)(12.5\ ft)(15\ ft) + (\pi)(12.5\ ft)^2 =$$

   $$1178.10\ ft^2 + 490.87\ ft^2 = 1668.97\ ft^2$$

STEP 2 Subtract area of 1 door:

(3 ft)(3 ft) = 9 ft²

1668.97 ft² - 9 ft² = 1659.97 ft²

STEP 3 Find total surface area of both tanks and 2 coats of paint:

2 coats x 2 tanks x 1659.97 ft² = 6639.88 ft²

STEP 4 Find total gallons of paint:

6639.88 ft² / (200 ft² / gallon) = 33.20 gallons

STEP 5 Find cost of paint:

$35/gallon x 33.20 gallons = $1,162.00

STEP 6 Calculate production time of the workers:

2 workers can paint 675 ft²/hour

6639.88 ft² / (675 ft²/hour) = 9.83 hours

9.83 hours / 2 workers = 4.9 hours each worker

Round to 5 hours labor for each worker

STEP 6 Calculate cost of the workers:

2 x $30/hour x 5 hours = $300

STEP 6 Calculate final materials and labor cost:

$1,162.00 + $300 = $1,462.00

Correct Answer: (A)

SOLUTION 6

Which of the following is NOT a concrete admixture:

- A. Air entraining
- B. Water reducing
- C. Corrosion inhibiting
- D. Shrinkage reducing
- E. Organic growth promoting
- F. Retarding (setting time reduction)
- G. Accelerating (setting time accelerator)

1. What type of problem is it?

   - Construction: Materials

2. What is the problem asking for?

   - Choose the best answer for which of the above is NOT a concrete admixture

3. Is there extraneous information?

   - No

4. What references or equations are needed?

   - Concrete admixtures

5. Solution:

"THINGS TO THINK ABOUT"
What if you were asked which admixture is appropriate for use in cold weather?

"THINGS TO REMEMBER"
When bubbling your answer, make sure you are referring to the correct choice
(see Problem 6, Choice B)

A) Air entering admixtures reduces strength but provides properties that perform well in freeze – thaw conditions.

B) Water reducing admixtures increase workability without increasing water / cement ratio; increasing water / cement Ratio reduce strength.

C) Corrosion inhibiting admixtures reduce rebar corrosion within the concrete.

D) Shrinkage reducing admixtures reduces cracking due to shrinkage.

E) Organic material should not be present in concrete.

F) Retarding admixtures reduces setting time.

G) Accelerating admixtures expedite setting time.

Correct Answer: (B)

SOLUTION 7

Find the correct moment diagram for the loading shown below.

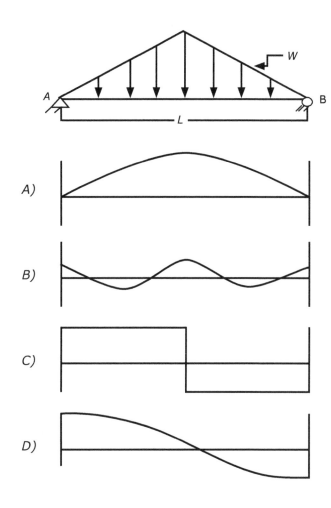

1. What type of problem is it?

   ■ Structural: Structural Mechanics

2. What is the problem asking for?

   ■ Moment diagram given loading

3. Is there extraneous information?

   ■ No

4. What references or equations are needed?

   ■ Shear and Moment diagrams

5. Solution:

STEP 1 Find the shear diagram:

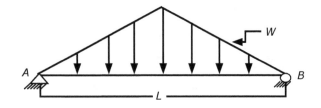

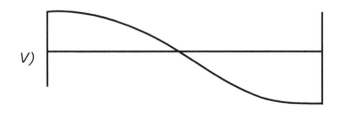

V)

"THINGS TO THINK ABOUT"

What if you were asked the equation for the maximum shear force at A and B?

$$V_{max} = \pm \frac{WL}{4}$$

STEP 2 Find the moment diagram, using the shear diagram:

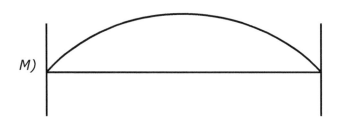

V)

M)

"THINGS TO REMEMBER"

When bubbling your answer, make sure you are referring to the correct choice
(see Problem 7, Choice B)

"THINGS TO THINK ABOUT"

What if you were asked the equation for the maximum moment force at midspan?

$$M_{max} = \frac{WL^2}{12}$$

Correct Answer: (B)

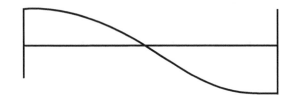

The moisture content for the fine and coarse aggregate for a concrete mix are 5.4% and 1.9%, respectively. The saturated surface dry (*SSD*) specific gravity (*SG*) for the fine and coarse aggregate are 2.69 and 2.63, respectively. The saturated surface dry (*SSD*) moisture content (ω_{SSD}) for the fine and coarse aggregate are 1.5% and 0.9%, respectively. The volume of a particular batch of concrete is 2.35 cf. Sulphates are not present where the concrete will be installed. The following table shows the weights of several crucial ingredients for the total concrete mix:

Weight of Water (lb)	Weight of Cement (lb)	Weight of Course Aggregate (lb)	Weight of Fine Aggregate (lb)
35.6	57.2	145.4	109.5

Find the water/cement ratio (w/c) for the concrete mix batch.

1. What type of problem is it?
   - Construction: Materials

2. What is the problem asking for?

"THINGS TO THINK ABOUT"
What does SSD mean? Why do we need to adjust to SSD?

   - The question is asking for the water/cement ratio (*w/c*) for the concrete mix batch

3. Is there extraneous information?

   - You do not need to know that sulphates are not present where the concrete will be installed to answer the question

"THINGS TO THINK ABOUT"
How does adding more water to the mix change the properties of the concrete?

   - You do not need to know the volume of the batch of concrete

   - The saturated surface dry (*SSD*) specific gravity (*SG*) for the fine and coarse aggregate amount is not needed to answer the question

4. What references or equations are needed?

- Concrete/cement mix

- The following equation is used for adjustment of the weight of the fine aggregate (*FA*) and coarse aggregate (*CA*) to saturated surface dry (*SSD*):

$$W_{SSD} = \left(\frac{W}{1 + (\omega)} \right)(1 + \omega_{SSD})$$

- The following equation is used for water/cement ratio (w/c):

$$\text{w/c} = \left(\frac{Weight\ of\ Water}{Weight\ of\ Cement} \right)$$

5. Solution:

STEP 1 Make a quick table of values from the problem statement, as shown below (note that SG is not needed, so those values do not need to go into your table):

	Moisture Content (ω)	SSD Moisture Content (ω_{SSD})
Fine Aggregate (FA)	0.054	0.015
Coarse Aggregate (CA)	0.019	0.009

STEP 2 Find the total weight of the mix without SSD adjustment:

35.6 lbs + 57.2 lbs + 145.4 lbs + 109.5 lbs = 347.7 lbs

STEP 3 Adjust of the weight of the fine aggregate (*FA*) to saturated surface dry (*SSD*) (Note that no adjustment is needed for the cement):

Fine Aggregate (FA):

$$W_{SSD} = \left(\frac{W}{1+(\omega)}\right)(1+\omega_{SSD}) = \left(\frac{109.5}{1+(0.054)}\right)(1+0.015) = 105.45 \ lbs$$

STEP 4 Adjust of the weight of the coarse aggregate (*CA*) to saturated surface dry (*SSD*) (Note that no adjustment is needed for the cement):

Coarse Aggregate (CA):

$$W_{SSD} = \left(\frac{W}{1+(\omega)}\right)(1+\omega_{SSD}) = \left(\frac{145.4}{1+(0.019)}\right)(1+0.009) = 143.97 \ lbs$$

STEP 5 Find the total weight of the water, using the total weight of the mix without SSD adjustment, cement weight, and the SSD adjusted weights of the fine and coarse aggregates (assuming the only components of the concrete mix are cement, water, and aggregate):

347.7 lbs – 105.45 lbs – 143.97 lbs – 57.2 lbs = 41.08 lbs

STEP 6 Find the water/cement ratio (*w/c*):

$$w/c = \left(\frac{Weight \ of \ Water}{Weight \ of \ Cement}\right) = \left(\frac{41.08 \ lbs}{57.2 \ lbs}\right)\left(\frac{\frac{1 \ gallon}{8.34 \ lbs}}{\frac{1 \ sack \ of \ cement}{94 \ lbs}}\right) = 8.09 \ \frac{gallons}{sack}$$

Correct Answer: (B)

```
"THINGS TO REMEMBER"
1 gallon = 8.34 lbs
1 ft³ of water = 62.4 lbs
1 sack of cement is
approximately 94 lbs
```

SOLUTION 9

A 22 ft beam has a pinned connection at one end (*at A*) and a roller at the other end (*at B*), as shown below. The beam is subjected to a uniform load (W) of 0.6 kip/ft and a point load (P) of 5.5 kips that is 5 ft from the left end. The beam is 2 x 6 Douglas Fir. Find the location of maximum moment (M_{max}).

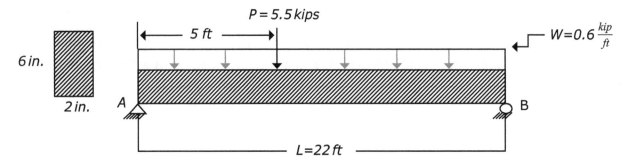

1. What type of problem is it?

   ■ Structural: Structural Mechanics

2. What is the problem asking for?

   ■ Location of maximum moment

3. Is there extraneous information?

   ■ Dimensions of cross-section not needed

   ■ Type of material (Douglas Fir) not needed

4. What references or equations are needed?

   ■ Structural design

5. Solution:

   STEP 1 Find equivalent force on the beam:

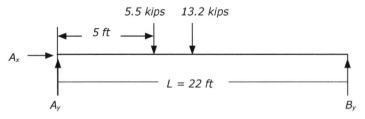

STEP 2 Find Reaction at B by taking sum of moments about point A:

$$\sum M_A = O; \ B_y \ (22ft) = (13.2 \ kips)(11ft) + (5.5 \ kips)(5ft)$$

$$\therefore \ B_y = 7.85 \ kips$$

STEP 3 Find reaction at A by taking sum of forces in the y-direction and x-direction:

$$\sum F_{y-direction} = O; \ A_y = 5.5 \ kips + 13.2 \ kips - 7.85 \ kips$$

$$\therefore \ Ay = 10.85 \ kips$$

$$\sum F_{x-direction} = O; \ A_x = 0$$

STEP 4 Draw shear diagram to find maximum moment (M_{max}):

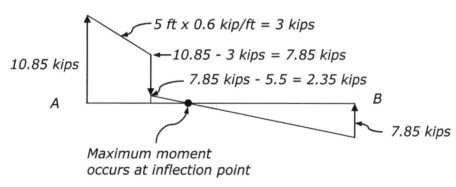

STEP 5 Use similar triangles to find location of maximum moment.

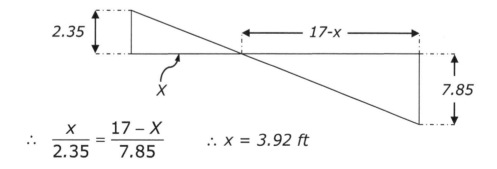

$$\therefore \ \frac{x}{2.35} = \frac{17 - X}{7.85} \qquad \therefore \ x = 3.92 \ ft$$

∴Distance to point load plus additional distance to inflection point is 5 ft + 3.92 ft = 8.92 ft from Point A

∴Maximum moment (M_{max}) is at 8.92 ft from Point A

Correct Answer: (D)

SOLUTION 10

35 full-time employees (40 hrs/week and 2000 hrs/yr) and 20 half-time employees (20 hrs/week and 1000 hrs/yr) work for a concrete construction company. 4 half-time employees were treated with first aid for minor cuts and scrapes, 3 half-time employees had injuries resulting in "light duty", 3 full-time employees were medically treated, but had no time-off or "light duty", and 2 full-time employees were injured and took time-off. Find the OSHA Recordable Incident Rate (*IR*).

"THINGS TO THINK ABOUT"
What if you were asked for the DART rate?

---

1. What type of problem is it?

   ■ Construction: Site Development

2. What is the problem asking for?

   ■ The question is asking for the OSHA Recordable Incident Rate, or Incident Rate (*IR*)

3. Is there extraneous information?

   ■ The number of full-time and half-time employees with injuries is not needed to answer the question

   ■ The amount of time the employees took time-off is not needed to answer the question

---

"THINGS TO THINK ABOUT"
What if you were asked for the Severity Rate?

4. What references or equations are needed?

- OSHA Reference Manual

- The following equation is used for the OSHA Recordable Incident Rate (IR):

$$IR = \left(\frac{Number\ of\ OSHA\ Recordable\ Cases\ x\ 2,000\ hrs/yr}{Number\ of\ Employees\ Labor\ Hours\ Worked} \right)(100\%)$$

- OSHA Recordable Cases:

  a. Any lost-time injuries

  b. Medical injuries resulting in "light duty".

  c. Medical treatment injuries (no "light duty", no lost-time)

- Cases that are not recordable:

  a. Minor injuries treated with first-aid only

5. Solution:

- The Number of OSHA Recordable Cases = 8

- Number of Employee Labor Hours Worked = (35 x 2000 hrs/yr) + (20 x 1000 hrs/yr)

- Enter these values into the equation for IR as shown below

$$IR = \left(\frac{8\ x\ 2,000\ hrs/yr}{(35 \times 2,000\ hrs/yr) + (20 \times 1,000\ hrs/yr)} \right)(100\%) = 17.78\%$$

Correct Answer: (B)

SOLUTION 11

A university transportation engineering class conducted a traffic count of a major arterial road during evening rush-hour traffic. Find the peak hour factor (*PHF*) using the given vehicle volume and time interval.

Time Interval	Volume (Vehicles)	Time Interval	Volume (Vehicles)
4:30pm – 4:45pm	390	5:30pm – 5:45pm	575
4:45pm – 5:00pm	410	5:45pm – 6:00pm	560
5:00pm – 5:15pm	540	6:00pm – 6:15pm	420
5:15pm – 5:30pm	570	6:15pm – 6:30pm	400

1. What type of problem is it?

   ▪ Transportation: Geometrics

2. What is the problem asking for?

   ▪ Find the peak hour factor (*PHF*)

3. Is there extraneous information?

   ▪ The fact that a class performed the study is not needed

   ▪ The fact that it is a major arterial road is not needed

4. What references or equations are needed?

   ▪ $PHF = \dfrac{maximum\ hourly\ volume}{4\ (peak\ 15-minute\ volume)} = \dfrac{V_{vph,max}}{4\left(V_{15-min,peak}\right)}$

5. Solution:

STEP 1 Find $V_{vph,max}$

- Find the largest four consecutive numbers and add them up (this is the largest number of vehicles during a one-hour period).

$$\therefore 540 + 570 + 575 + 560 = 2245 \ vehicles$$

STEP 2 Find $V_{15-min,peak}$

- The peak 15-minute volume occurs between 5:30pm and 5:45pm and equals 575 vehicles.

STEP 3 $\therefore PHF = \dfrac{maximum\ hourly\ volume}{4\ (peak\ 15-minute\ volume)} = \dfrac{V_{vph,max}}{4\ (V_{15-min,peak})}$

$\therefore PHF = 0.98$

Correct Answer: (C)

A concrete "gravity" retaining wall with normal weight concrete ($\Upsilon_c=155\ lb/ft^3$) holds soil backfill ($\Upsilon_s=120\ lb/ft^3$), as shown below. The wall is 10 ft in height (H = 10 ft), 3 ft thick at the top, and 5 ft thick at the bottom, and 20 ft long. $\beta=0°$, $\alpha=0°$, $\varnothing=36°$, and the cohesion of the soil is c=0°. Ignore the exterior vertical wall friction, $\delta=0°$. However, assume the exterior friction under the base of the wall is $\delta_{base}=26°$. The water table is 2 ft from the top of the wall (h = 2 ft). Find the Factor of Safety for sliding (FOS) for the retaining wall, using the Rankine Theory for active earth pressure.

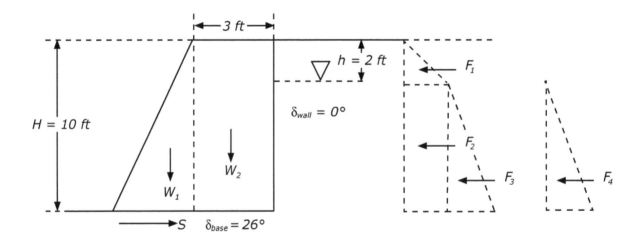

1. What type of problem is it?

 ■ Geotechnical: Soil Mechanics

2. What is the problem asking for?

 ■ The question is asking for the Factor of Safety for sliding (FOS) for the retaining wall

3. Is there extraneous information?

 ■ The length of the wall is not needed to answer the question

 ■ The value of cohesion (c) is not needed to answer the question

4. What references or equations are needed?

- Reference Lateral Earth Pressure

- The following equation is used for active earth pressure coefficient:

$$K_\alpha = \cfrac{\cos^2(\emptyset - \alpha)}{\cos^2(\alpha)\cos(\delta+\alpha)\left(1+\sqrt{\cfrac{\sin(\emptyset+\delta)\sin(\emptyset-\beta)}{\cos(\delta+\alpha)\cos(\alpha-\beta)}}\right)^2}$$

- The following equation is used for pressure within soil:
$p_v = (\rho g)H = \Upsilon H$

- The following equation is used for active pressure within granular soil: $P_a = K_a P_v = K_a(\Upsilon H)$

- The following equation is used for total active resultant force within granular soil:

$$F_a = \frac{1}{2}P_a H = \frac{1}{2}(K_a P_v)H = \frac{1}{2}(K_a \gamma H)H = \frac{1}{2}K_a \gamma H^2$$

- The following equation is used for the sliding force:
$S = W_{total}\tan(\delta_{base})$

- The following equation is used for FOS for sliding:

$$FOS = \frac{S}{\sum F}$$

5. Solution:

STEP 1 Draw the diagram above, including all forces.
(Note that forces of triangular distribution act at the H/3 from the base.)

STEP 2 When $\alpha = 90$, $\beta = 0$, and $\delta_{wall} = 0$, then:

$$K_a = \tan^2\left(45° - \frac{\emptyset}{2}\right)$$

"TIPS"
Know the meaning of these symbols (e.g., β is the angle of the soil backfill to the horizontal)

STEP 3 Find K_a:

The friction angle between the soil and the concrete at the vertical wall is $\delta_{wall} = 0°$. The friction angle between the soil and the concrete at the base of the wall is $\delta_{base} = 26°$. The angle of the soil backfill to the horizontal is $\beta = 0°$. The angle of the vertical surface of the retaining wall to the horizontal is $\theta = 90°$. The angle of internal friction is $\emptyset = 36°$. Enter these values, as degrees, into the equation for K_α, as shown below.

$$K_\alpha = \left(\frac{\cos^2(36° - 0)}{\cos^2(0)\cos(0+0)\left(1 + \sqrt{\frac{\sin(36°+0)\sin(36°-0)}{\cos(0+0)\cos(0-0)}}\right)^2} \right) = \tan^2\left(45° - \frac{36°}{2}\right) = 0.260$$

"THINGS TO REMEMBER"
$\cos(\beta) = 0°$, for $\beta = 0°$
$\sin(\alpha) = 1$, for $\alpha = 90°$
$\sin(90 + \beta) = \cos(\beta)$
$\sin(90 + \emptyset) = \cos(\emptyset)$
$\sin(90 - \delta) = \cos(\delta)$

Therefore, $\beta = 0°$, and $\delta = 0°$ (i.e., when the soil backfill is horizontal and the wall surface is vertical), then:

$$K_\alpha = \tan^2\left(45° - \frac{\emptyset}{2}\right)$$

STEP 4 Find the equations for vertical pressures within the soil:

  a. Pressure above the water table $= K_a \gamma_s (h)$

  b. Pressure below the water table $= K_a \gamma_s h_1 + K_a (\gamma_s - \gamma_{water})(H - h_1)$

  c. Water pressure $= \gamma_w (H - h_1)$

STEP 5 Find all resultant forces (combine with equations for vertical pressures within the soil):

  a. Resultant force above the water table:

$$F_1 = \frac{1}{2}K_a(\gamma_s)(h_1)^2 = \frac{1}{2}(0.260)\left(120\frac{lb}{ft^3}\right)(2ft)^2 = 62.4\frac{lb}{ft}$$

  b. Resultant forces below the water table:

$$F_2 = K_a(\gamma_s)(h)(H - h) = (0.260)\left(120\frac{lb}{ft^3}\right)(2ft)(8ft) = 499.2\frac{lb}{ft}$$

$$F_3 = \frac{1}{2}K_a(\gamma_s - \gamma_{water})(H - h)^2 = \frac{1}{2}(0.260)\left(120\frac{lb}{ft^3} - 62.4\frac{lb}{ft^3}\right)(8ft)^2 = 479.23\frac{lb}{ft}$$

c. Water pressure force:

$$F_4 = \frac{1}{2}\Upsilon_w\left(H - h_1\right)^2 = \frac{1}{2}\left(62.4\,\frac{lb}{ft^3}\right)\left(8ft\right)^2 = 1996.8\,\frac{lb}{ft}$$

STEP 6 Find W1 and W2:

$$W_1 = \left(\frac{1}{2}\right)\left(2\,ft\right)\left(10\,ft\right)\left(155\,\frac{lb}{ft^3}\right) = 1550\,\frac{lb}{ft}$$

$$W_2 = \left(3\,ft\right)\left(10\,ft\right)\left(155\,\frac{lb}{ft^3}\right) = 4650\,\frac{lb}{ft}$$

STEP 7 Find sliding force:

The sliding force is

$$S = \left(W_1 + W_2\right)tan\left(\delta_{base}\right) = \left(1550\,\frac{lb}{ft} + 4650\,\frac{lb}{ft}\right)\left(tan\left(26°\right)\right) = 3023.94\,\frac{lb}{ft}$$

STEP 8 Find FOS:

The FOS for sliding is

$$\frac{3023.94\,\frac{lb}{ft}}{499.2\,\frac{lb}{ft} + 62.4\,\frac{lb}{ft} + 479.23\,\frac{lb}{ft} + 1996.8\,\frac{lb}{ft}} = 0.995\ \textit{(not adequate)}$$

Correct Answer: (B)

"THINGS TO THINK ABOUT"
FOS for sliding should be greater
than 1.5.

SOLUTION 13

The lowest point of an overpass is required to be 24 ft above the high point of a crest curve, as shown. The distance between the PT and the PI is 400 ft. The elevation of the PI is 1100 ft. Find the elevation of the high point of the curve (Y_{HP}).

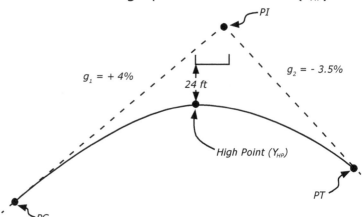

1. What type of problem is it?

    ■ Transportation: Geometrics

2. What is the problem asking for?

    ■ The elevation of the high point of the curve (Y_{HP})

3. Is there extraneous information?

    ■ The information requiring the overpass is not needed

4. What references or equations are needed?

    ■ The general equation for the high point of a crest vertical curve is:

    $$Y_{HP} = Y_{PC} - \frac{L(g_1)^2}{2(g_2 - g_1)}$$

    ■ The general equation for the elevation at PC is:

5. Solution:

    $$Y_{PC} = Y_{PI} \pm g_1 \frac{L}{2}$$

STEP 1 Find the elevation at PC:

$$Y_{PC} = 1100ft - (0.04)\left(\frac{(2)(400)}{2}\right) = 1,084 \ ft$$

STEP 2 Find the elevation at the high point (Y_{HP}):

$$Y_{HP} = 1084ft - \left(\frac{(2)(400)(0.04)^2}{2(-0.035-0.04)}\right) = 1,092.53 \ ft$$

Correct Answer: (C)

"THINGS TO REMEMBER"
For vertical curves, PI is located at L/2, this comes in handy for finding L if given PI or vice versa.

"THINGS TO REMEMBER"
L for a vertical curve is the horizontal length between the PC and the PT, not the length of the actual curve along the road.

SOLUTION 14

A concrete retaining wall with normal weight concrete ($\Upsilon_c=155$ lb/ft^3) holds soil backfill ($\Upsilon_s=120$ lb/ft^3), as shown below. The wall is 12 ft in height, 2 ft thick at the top, 3 ft thick at the bottom, and 20 ft long. The friction angle between the soil and the concrete is (δ)=21°. The angle of the soil backfill to the horizontal (β) is 1V:2H. The angle of the surface of the retaining wall to the vertical is (α)=15°. The angle of internal friction is ($\emptyset$)=31°. The cohesion of the soil is (c)=0°. Ignore friction under the wall. Draw the angles into the diagram below and find the active earth pressure coefficient.

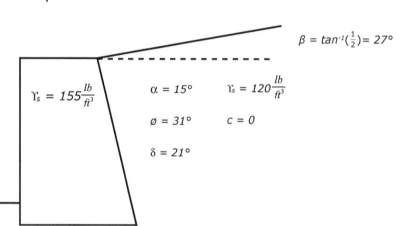

$$\Upsilon_s = 155\frac{lb}{ft^3}$$

$$\beta = tan^{-1}(\tfrac{1}{2}) = 27°$$

$\alpha = 15°$ $\Upsilon_s = 120\frac{lb}{ft^3}$

$\emptyset = 31°$ $c = 0$

$\delta = 21°$

1. What type of problem is it?

   ▪ Geotechnical: Soil mechanics

2. What is the problem asking for?

   ▪ The question is asking for K_a, the active earth pressure coefficient

3. Is there extraneous information?

   ▪ The length of the wall is not needed to answer the question

---

"THINGS TO THINK ABOUT"
What if you were asked for the total active resultant force?

"THINGS TO THINK ABOUT"
What if you were asked for "at rest" or "passive" earth pressure coefficient?

"THINGS TO THINK ABOUT"
For (saturated) clay: $\beta \leq \emptyset$
For granular soils (e.g.,sand): $C = 0$

■ The unit weight of the concrete and soil are not needed to answer the question

■ The dimensions of the footing and wall are not needed to answer the question

■ The value of cohesion (c) is not needed to answer the question

4. What references or equations are needed?

■ Reference Lateral Earth Pressure

■ Coulomb's equation is used for active earth pressure coefficient (only valid for $\beta \leq \emptyset$):

$$K_a = \left(\frac{\cos^2(\emptyset - a)}{\cos^2(a)\cos(\delta + a)\left(1 + \sqrt{\frac{\sin(\emptyset + \delta)\sin(\emptyset - \beta)}{\cos(\delta + a)\cos(a - \beta)}}\right)^2} \right)$$

> **"THINGS TO REMEMBER"**
> $\cos(\beta) = 0°$, for $\beta = 0°$
> $\sin(\alpha) = 1$, for $\alpha = 90°$
> $\sin(90 + \beta) = \cos(\beta)$
> $\sin(90 + \emptyset) = \cos(\emptyset)$
> $\sin(90 - \delta) = \cos(\delta)$
>
> Therefore, $\beta = 0°$, and $\delta = 0°$, (i.e., when the soil backfill is horizontal and the wall surface is vertical), then:
>
> $$K_a = \tan^2\left(45° - \frac{\emptyset}{2}\right)$$

5. Solution:

The friction angle between the soil and the concrete is $\delta = 21°$. The angle of the soil backfill to the horizontal is $\beta = \tan^{-1}(1/2) = 27°$ The angle of the vertical surface of the retaining wall to the horizontal is $\alpha = 15°$. The angle of internal friction is $\emptyset = 31°$. The cohesion of the soil is $c = 0$. Enter these values, as degrees, into the equation for K_a, as shown below.

$$K_a = \left(\frac{\cos^2(31° - 15°)}{\cos^2(15°)\cos(21° + 15°)\left(1 + \sqrt{\frac{\sin(31° + 21°)\sin(31° - 27°)}{\cos(21° + 15°)\cos(15° - 27°)}}\right)^2} \right) = 0.767$$

Correct Answer: (B)

SOLUTION 15

The lowest point of an overpass is required to be a minimum of 18 ft above point P on a vertical sag curve as shown. Find the minimum elevation of the overpass if station of the Point of Tangency (*PT*) is 17+00 and the station at a point on the curve (*P*) is 8+02. The distance between PC and PI is 525 ft. The elevation of PC is 255 ft, The elevation of PT is 400 ft.

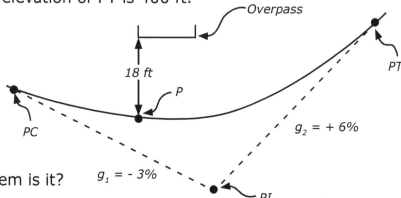

1. What type of problem is it?

    ■ Transportation: Geometrics

2. What is the problem asking for?

    ■ Minimum elevation of the overpass

3. Is there extraneous information?

    ■ No

4. What references or equations are needed?

    ■ The general equation for a vertical curve is:
    $Y_P = Y_{PC} + g_1 x + \dfrac{(g_2 - g_1)x^2}{2L}$, where x is the distance between PC and Point P.

    ■ The general equation for the elevation at PC is:
    $Y_{PC} = Y_{PI} \pm g_1 \dfrac{L}{2}$

    ■ The general equation for the elevation at PT is:
    $Y_{PT} = Y_{PI} \pm g_2 \dfrac{L}{2}$

    ■ The station at $PC(PC_{STA})$ the station at: $PT(PT_{STA})$ - L

    ∴ $PC_{STA} = PT_{STA} - L$

5. Solution:

STEP 1 Find PC_{STA}:

$$PC_{STA} = PT_{STA} - L$$

$$\therefore PC_{STA} = 1,700 \text{ ft} - [2(525 \text{ ft})] = 650 \text{ ft}$$

STEP 2 Find x (the distance between PC and Point P):

$$\therefore x = 802 \text{ ft} - 650 \text{ ft} = 152 \text{ ft}$$

STEP 3 Find the elevation at P:

$$Y_P = Y_{PC} + g_1 x + \frac{(g_2 - g_1)x^2}{2L}$$

$$\therefore Y_P = 225ft + (-0.03)(152ft) + \frac{(0.06 + 0.03)(152ft)^2}{2[2(525ft)]} = 251.43ft$$

STEP 4 Find the minimum elevation of the overpass:

251.43 ft + 18 ft = 269.43 ft

Correct Answer: (A)

The activity-on-arrow for a bridge project in Yellowstone National Park is shown. Find the early start (*ES*) and early finish (*EF*) for activity K.

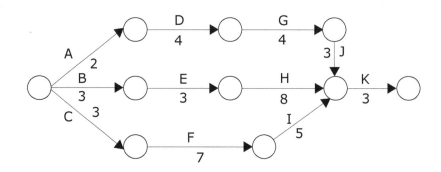

Activity	Duration
A	2
B	3
C	3
D	4

Activity	Duration
E	3
F	7
G	4
H	8

Activity	Duration
I	5
J	3
K	3

1. What type of problem is it?

   ■ Construction: Project Planning

2. What is the problem asking for?

   ■ Early Start (*ES*) and Early Finish (*EF*) of activity K

3. Is there extraneous information?

   ■ The fact that it is a bridge project in Yellowstone is not needed

4. References

   Project scheduling

5. Solution:

   ■ *ES = 15, EF = 18*   **Correct Answer: (C)**

SOLUTION 17

A vertical crest curve has a +2% and a -6% grade as shown below. The distance between the Point of Curvature (*PC*) and the Point of Intersection (*PI*) is 1,125 ft. The station of the Point of Tangency (*PT*) is 42+25. The station at a Point on the curve (*P*) is 28+75. The elevation at PC (Y_{PC}) is 100 ft. Find the slope at Point P (S_P).

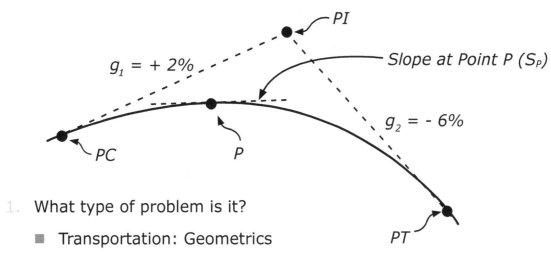

1. What type of problem is it?

   ■ Transportation: Geometrics

2. What is the problem asking for?

   ■ Slope at Point P *(S_P)*

3. Is there extraneous information?

   ■ No

4. What references or equations are needed?

   ■ The general equation for a vertical curve is:

   $$Y_P = Y_{PC} + g_1 x + \frac{(g_2 - g_1) x^2}{2L}$$, where x is the distance between PC and Point P.

   ■ Take the derivative of the general equation for a vertical curve to find the equation for the slope:

   $$S_P = g_1 + \frac{(g_2 - g_1) x}{L}$$, where x is the distance between PC and Point P.

   ■ The station a *PC* (PC_{STA}) = the station at *PT* (PT_{STA}) - L

   $$\therefore PC_{STA} = PT_{STA} - L$$

5. Solution:

1 Find PC_{STA}:

> **"THINGS TO REMEMBER"**
> For vertical curves, the length of the curve (L) is defined as $2(PI_{STA} - PC_{STA})$

$PC_{STA} = PT_{STA} - L$

$\therefore PC_{STA} = 4{,}225\ ft - [2(1{,}125)] = 1{,}975\ ft$

STEP 2 Find x (the distance between PC and Point P):

$x = 2{,}875\ ft - 1{,}975\ ft = 900\ ft$

STEP 3 Find the slope at Point P (S_P):

$\therefore S_P = 0.02 + \left(\dfrac{-0.06 - 0.02}{[2(1{,}125)]}\right)(900\ ft) = -0.012$

$\therefore S_P = -1.2\%$

Correct Answer: (A)

© 2015 | passthecivilPE.com

SOLUTION

18

Modular steel frame formwork is used for a retaining wall that is 11.5 ft in height. The temperature outside is 68°F. The temperature of the concrete is 75°F. The cement contains 44% fly ash and the concrete has density of 155 lb/ft³. The concrete will be placed at 7.5 ft/hr. Find the lateral pressure of the concrete on the formwork using ACI 347.

1. What type of problem is it?

   ▪ Construction: Means and Methods

2. What is the problem asking for?

   ▪ Lateral pressure of the concrete using ACI 347

3. Is there extraneous information?

   ▪ The temp outside is not needed

4. References ACI 347:

$$P_{wall} = C_w C_c \left(150 + 9000 \, R/t\right) \left(R \leq 7 \frac{ft}{hr}\right)$$

$$P_{wall} = C_w C_c \left(150 + \frac{43,400}{T} + \frac{2,800 \, R}{T}\right) \left(R > 7 \frac{ft}{hr}\right)$$

5. Solution:

   STEP 1 $R > 7ft \frac{ft}{hr} \therefore use \; P_{wall} = C_w C_c \left(150 + \frac{43,400}{T} + \frac{2,800 \, R}{T}\right)$

   STEP 2 Use *ACI 347* to find C_w and C_C:

   $$\therefore C_w = \frac{155}{145} = 1.07, \; since \; Density = 155 \; pcf$$

   $$\therefore C_c = 1.4, \; since \; FLYASH > 40\%$$

STEP 3 Temperature of the concrete is 75°F:

$$\therefore T = 75°F$$

STEP 4 Find the lateral pressure of the concrete:

$$P_{wall} = (1.07)(1.4)\left(150 + \frac{43,400}{75°F} + \frac{2,800\left(7.5\frac{ft}{hr}\right)}{75°F}\right) = 1,511\frac{lb}{ft^2}$$

STEP 5 Check minimum pressure requirements:

$$P_{min} = 600C_w = 600 (1.07) = 642 \text{ lb/ft}^2$$

Correct Answer: (C)

SOLUTION 19

Which of the following is not one of the primary intentions and purpose of the Occupational Safety and Health Administration (OSHA) regulations? Choose the best answer.

A) Any worker (with an employer regulated under the jurisdiction of OSHA) may file a complaint to have OSHA inspect their workplace if they believe that their employer is not following OSHA standards, or there are serious hazards in their workplace.

B) OSHA only applies to construction workers and work related to construction.

C) The Occupational Safety and Health Act of 1970 (OSH Act), which led to the creation of OSHA, requires employers to provide their employees with working conditions that are free of known dangers.

D) OSHA regulates construction worker safety separately from its general industry, maritime, and agricultural requirements, and provides construction industry regulations which can be found in the standard publication titled 29 CFR 1926.

1. What type of problem is it?

   - Construction: Site Development

2. What is the problem asking for?

   - Best answer for false statement regarding OSHA regulations

3. Is there extraneous information?

   - No

4. What references or equations are needed?

   - Reference Occupational Safety and Health Act of 1970 (OSH Act)

   - Reference OSHA 29 CFR 1926

5. Solution:

OSHA's purpose is to ensure a safe and healthy workplace and working conditions for every working man and woman in the United State of America, by setting and enforcing standards and by providing assistance, information, and education; this includes all workplaces, not only construction related jobs.

Correct Answer: (C)

SOLUTION 20

A temporary structure has been built on a construction site in Frederick, Maryland. The structure will be in place for approximately 1.5 years. The design wind speed used for permanent structures adjacent to the temporary structure is 85 mph. Find the lateral wind pressure used to design the temporary structure using ASCE 37.

1. What type of problem is it?

   - Construction: Means and Methods

2. What is the problem asking for?

   - Lateral pressure on temporary structure

3. Is there extraneous information?

   - The fact that it is in Frederick, MD is not needed

4. References:

   - ASCE 37: $P_{wind} = 0.00256\ V_c^2$ (where V_c = Factored wind speed used during construction)

   - From ASCE 7:

Construction period	Design Wind Speed
< 6 weeks	0.75
> 6 weeks, < 12 months	0.80
> 12 months, < 24 months	0.85
> 24 months, < 60 months	0.90

5. Solution:

   STEP 1 Wind speed factor is 0.85 (12 months < construction period < 24 months)

   $V_c = (0.85)(85mph) = 72.25\ mph$

   STEP 2 $P_{wind} = (0.00256)(72.25)^2 = 13.36\ psf$ **Correct Answer: (D)**

SOLUTION 21

Which of the following is false regarding open channel flow? Choose the best answer.

A) In general, the most efficient open channel cross section will have a maximized hydraulic radius, R, and a minimized wetted perimeter, P.

B) The usual minimum permissible velocity of an open channel, including closed-conduit pipe networks flowing under the influence of gravity only, is the lowest velocity that prevents sediment deposit.

C) The liquid sheet Nappe of a rectangular weir decreases in width as it falls if the weir opening width is less than the channel width.

D) Laminar flow occurs at high Reynolds numbers, and turbulent flow occurs at low Reynolds numbers.

1. What type of problem is it?

   ■ Water Resources and Environmental: Hydraulics and Hydrology

2. What is the problem asking for?

   ■ Best answer for false statement regarding open channel flow

3. Is there extraneous information?

   ■ No

4. What references or equations are needed?

   ■ Reference water resources engineering and hydrology for open channel flow

5. Solution:

   Laminar flow occurs at low Reynolds numbers with dominant viscous forces, and characterized by smooth, constant fluid motion; turbulent flow occurs at high Reynolds numbers, dominated by internal forces, and characterized by random and instable fluid motion.

Correct Answer: (C)

22

SOLUTION

A new water tower concrete foundation slab specification requires modified proctor testing to establish the soil compaction parameters. Five tests were performed with the following results at the laboratory.

Test	Weight	Moisture Content (ω)
1	5.72 lb	8.3%
2	5.96 lb	8.9%
3	6.15 lb	9.3%
4	6.13 lb	9.5%
5	5.89 lb	10.3%

At the construction site, 0.02 ft³ of completed soil is tested and has a wet weight of 2.4 lb and dry weight of 2.1 lb. Find the percentage of compaction for the in-situ soil.

1. What type of problem is it?

   ■ Geotechnical: Materials

2. What is the problem asking for?

   ■ Percentage of compaction for the in-situ soil

3. Is there extraneous information?

   ■ The fact that compaction soil is for a water tower concrete foundation slab is not needed

4. What references or equations are needed?

   ■ Reference soil properties and testing

"THINGS TO REMEMBER"

The volume of standard and modified proctor test mold is

$$V_{mold} = \frac{1}{30} ft^3$$

---

5. Solution:

STEP 1 Find the dry density $\Upsilon_{d\ in\text{-}situ}$ of the in-situ soil

$$Y_{d\ in\text{-}situ} = \frac{2.1 lb}{0.02 ft^3} = 105 \frac{lb}{ft^3}$$

Find the maximum dry density (Υ_{dmax}) using the laboratory results.

Test	Dry Density $\left[Y_d = \dfrac{w}{(1+\omega)V_{mold}} \right]$
1	$\dfrac{5.72 lb}{(1.083)\left(\dfrac{1}{30} ft^3\right)} = 158.45 \dfrac{lb}{ft^3}$
2	$\dfrac{5.96 lb}{(1.089)\left(\dfrac{1}{30} ft^3\right)} = 164.19 \dfrac{lb}{ft^3}$
3	$\dfrac{6.15 lb}{(1.093)\left(\dfrac{1}{30} ft^3\right)} = 168.80 \dfrac{lb}{ft^3}$
4	$\dfrac{6.13 lb}{(1.095)\left(\dfrac{1}{30} ft^3\right)} = 167.95 \dfrac{lb}{ft^3}$
5	$\dfrac{5.89 lb}{(1.103)\left(\dfrac{1}{30} ft^3\right)} = 160.20 \dfrac{lb}{ft^3}$

"THINGS TO REMEMBER"

Soil with optimum water content does not require as much compaction to reach specified soil compaction

STEP 3 Graph values found in Step 2 to find maximum dry density (Υ_{dmax}) and optimum moisture content (ω_{opt})

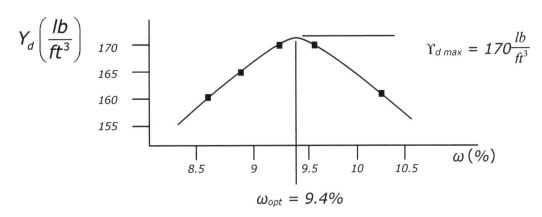

$$\Upsilon_{d\ max} = 170\frac{lb}{ft^3}$$

$$\omega_{opt} = 9.4\%$$

$$\therefore Y_{d\ max} = 170\frac{lb}{ft^3} \quad and \quad \omega_{opt} = 9.4\%$$

STEP 4 Find percentage of compaction:

$$\frac{Y_d in-situ}{Y_{d\ max}} = \frac{105\frac{lb}{ft^3}}{170\frac{lb}{ft^3}} = 0.618 = 61.8\%$$

(note: needs more compaction to comply with specifications)

"THINGS TO THINK ABOUT"
What if you were asked if the soil needed more or less water?

Correct Answer: (B)

SOLUTION

23

Due to a setback issue near the city of Bakersfield, California, an open channel section has the dimensions shown below. The slope of the concrete channel is 0.0004. Find the hydraulic radius (*R*) of the channel section.

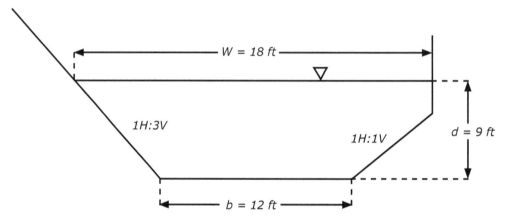

1. What type of problem is it?

   ◾ Water Resources and Environmental: Hydraulics and Hydrology

2. What is the problem asking for?

   ◾ The Hydraulic radius (*R*) of the channel section

3. Is there extraneous information?

   ◾ The slope is not needed

   ◾ The fact that the channel is concrete is not needed

   ◾ The fact that the channel has a setback issue

   ◾ The fact that the channel is in Bakersfield, California

4. What references or equations are needed?

   ◾ Hydraulic parameters of channel sections:

   Hydraulic radius = $\dfrac{Area\ of\ water\ from}{wetted\ parameter}$

   $$\left(R = \frac{A}{P} \right)$$

5. Solution:

STEP 1 Find area by breaking into sections:

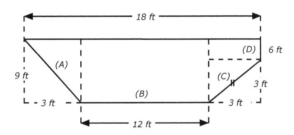

Section	Area (ft²)
(1)	$\frac{1}{2}(3)(9) = 13.5$
(2)	$(12)(9) = 108$
(3)	$(6)(3) = 18$
(4)	$\frac{1}{2}(3)(3) = 4.5$
Total	144

STEP 2 Find wetted parameters (i.e., surface of the channel in contact with water):

Perimeter Length (A) = $\sqrt{9^2 + 3^2}$ = 9.49 ft

Perimeter Length (B) = 12 ft

Perimeter Length (C) = $\sqrt{3^2 + 3^2}$ = 4.24 ft

Perimeter Length (D) = 6 ft

Total Perimeter Length = 9.49 ft + 12 ft + 4.24 ft + 6 ft = 31.73 ft

STEP 3 Find the hydraulic Radius:

$$R = \frac{A}{P} = \frac{144 \; ft^2}{31.73 \; ft} = 4.54 \; ft$$

"THINGS TO THINK ABOUT"
Slope and material type would be required if using Mannings Equation

Correct Answer: (C)

SOLUTION 24

Choose the answer below that is false:

A) The difference between liquid limit (*LL*) and the plastic limit (*PL*) is defined as the plasticity index (*PI*).

B) When the weight of water equals the weight of the dry soil (i.e., $\omega = 100\%$) in a soil sample the liquid limit (*LL*) is 100.

C) The plastic limit of a soil sample is attained when a soil sample that is rolled in to an 1/8 in. diameter thread begins to crumble.

D) Atterberg limit tests can be applied to other construction materials such as cement mixtures and certain asphaltic materials.

1. What type of problem is it?

 ■ Geotechnical: Materials

2. What is the problem asking for?

 ■ Find false statement

3. Is there extraneous information?

 ■ None

4. What references or equations are needed?

 ■ Geotechnical soils classifications

 ■ Atterberg limit tests

5. Solution:

 ■ The difference between liquid limit *(LL)* and the plastic limit *(PL)* is defined as the plasticity index *(PI)*.

 ■ When the weight of water equals the weight of the dry soil (i.e., $\omega = 100\%$) in a soil sample the liquid limit *(LL)* is 100.

 ■ The plastic limit of a soil sample is attained when a soil sample that is rolled in to a 1/8 in. diameter thread begins to crumble.

 ■ Atterberg limit tests can only be applied to soils.

Correct Answer: (B)

SOLUTION 25

A pump is used to dewater the soil adjacent to an elevator pit. The top of the concrete slab at the base of the elevator pit is 4 ft under the water table. The concrete slab is 3 ft thick. The walls of the elevator pit are concrete with blindside waterproofing and wood lagging with steel soldier piles. The water is pumped to a basin 20 ft above the top of the concrete slab. The specifications require the water table to be at least 2 ft below the bottom of slab. The pipe has a diameter of 6 in., length of 200 ft, and a Darcy Friction Factor of 0.04. The pump efficiency is 85% and the flow rate is 100 gallons per minute. Find the horsepower required for the pump.

1. What type of problem is it?

   - Water Resources and Environmental: Hydraulics and Hydrology

2. What is the problem asking for?

   - The horsepower required for the pump

3. Is there extraneous information?

   - The fact that it is an elevator pit or how the walls are constructed is not needed

4. What references or equations are needed?

   - Bernoulli's equation for energy conservation

   - Darcy equation (or Darcy – Weisbach equation) $h_f = \dfrac{fLv^2}{2Dg}$, which is the energy loss due to friction

   - *Pump Head = basin head $+ h_f$*

   - *Horse Power = $\dfrac{\gamma_w Q \left(pump\ head\right)}{550 \left(efficiency\right)}$*

5. Solution:

STEP 1 Find energy loss due to friction:

$$h_f = \frac{fLv^2}{2Dg}, \text{ where } f = 0.04, g = 32.2\frac{ft}{s^2}, L = 200\,ft$$

$$v = \frac{Q}{A} = \frac{100\,gpm}{\frac{\pi}{4}D^2} = \frac{100\,gpm}{\frac{\pi}{4}\left(\frac{6}{12}\right)^2} = 509.30\,\frac{gpm}{ft^2}\left(\frac{1\,ft^3}{7.4805_{gal}}\right)\left(\frac{1\,min}{60\,sec}\right) = 1.135\,ft/s$$

$$\therefore h_f = \frac{(0.04)(200\,ft)(1.135\,ft/s)^2}{2\left(\frac{6}{12}\right)(32.2\,ft/s^2)} = 0.32$$

STEP 2 Find basin head:

$$\therefore 20\,ft_{basin\,above\,slab} + 3\,ft_{slab\,thickness} + 2\,ft_{required\,depth\,of\,water\,below\,slab} = 25\,ft$$

STEP 3 Find the pump head:

Pump Head = Basin Head + h_f

$$\therefore 25\,ft + 0.32 = 25.32$$

STEP 4 Find horsepower:

$$\therefore Horsepower = \frac{(62.4\,pcf)(0.223\,ft^3/s)(25.32)}{550(0.85)} = 0.754\,hp$$

Correct Answer: (D)

SOLUTION 26

The specification for an airport access road in Sacramento, California, requires standard proctor compaction. The wet soil sample is 0.29 lbs and oven dried soil sample is 0.25 lbs. The wet soil sample and mold weighs 13.2 lbs and the empty mold weighs 9.5 lbs. Find the in-situ dry unit weight (Υ_d) of the soil.

1. What type of problem is it?

 - Geotechnical: Materials

2. What is the problem asking for?

 - In-situ unit weight (Υ_d) of the soil

"THINGS TO REMEMBER"

The volume of standard and modified proctor test mold is

$$V_{mold} = \frac{1}{30} ft^3$$

3. Is there extraneous information?

 - The fact that it is an airport access road is not needed, or that the project is in Sacramento

4. What references or equations are needed?

 - Reference soil standard permeation tests

 - Reference soil indexing formulas

5. Solution:

STEP 1 Find $\Upsilon_{wet} = \Upsilon_{total}$

$$Y_{wet} = Y_{total} = \frac{W_{soil+mold} - W_{mold}}{V_{mold}} = \frac{13.2lb - 9.5lbs}{\frac{1}{30} ft^3} = 111 \frac{lb}{ft^3}$$

STEP 2 Find In-situ soil moisture content $\omega_{in\text{-}situ}$

$$\omega_{in-situ} = \frac{\omega_{wet\ soil} - \omega_{dry\ soil}}{\omega_{dry\ soil}} = \frac{0.29 - 0.25}{0.25} = 0.16 = 16\%$$

STEP 3 Find dry density Υ_d

$$Y_d = \frac{Y_{total}}{1 + \omega_{in-situ}} = \frac{111 \frac{lb}{ft^3}}{1.16} = 95.69 \frac{lb}{ft^3}$$

Correct Answer: (B)

SOLUTION 27

Which of the following is false regarding slope stability at a construction site? Choose the best answer.

A) In general, unshored sides of soil should not be steeper than their in-situ angle of repose.

B) The most probable type of slip-plane for a trench in soft mud with walls sloped at 55° is "Toe Circle".

C) Slope failures typically occur in the shape of a sliding rectangle, where the depth of the sliding soil is approximately the same thickness at the top, middle, and toe of the slide.

D) In general, if the forces available within the soil to resist movement are greater than the forces driving movement, the slope is considered stable.

1. What type of problem is it?

   ■ Geotechnical: Soil Mechanics

2. What is the problem asking for?

   ■ Best answer for false statement regarding slope stability

3. Is there extraneous information?

   ■ No

4. What references or equations are needed?

   ■ Reference geotechnical engineering, soil and slope stability

5. Solution:

The surfaces of sliding for most slope failures have been observed to follow approximately the arc of a circle.

Correct Answer: (C)

SOLUTION 28

The mass diagram is shown below for a section of a new highway project in Pittsburgh, Pennsylvania. Choose the best answer below for which is true, given the mass diagram shown.

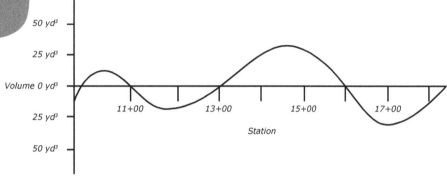

A) The section between Station 13+00 and 16+00 represents a fill operation.

B) The total grading operation is balanced.

C) The section between Station 11+00 and 16+00 represents the "free-haul distance".

D) The section at Station 17+00 represents a transition between fill and cut.

1. What type of problem is it?

   ■ Construction: Site Development

2. What is the problem asking for?

   ■ Choose the best answer below, given the mass diagram shown

3. Is there extraneous information?

   ■ The fact that it is a section of a new highway project Pittsburg, Pensylvania is not needed to answer the question

---

"THINGS TO THINK ABOUT"

What if the problem asked you to draw the original ground profile?

---

4. What references or equations are needed?

   ■ Reference properties of earth mass diagram

5. Solution:

A) The section between Station 13+00 and 16+00 does NOT represent a fill operation. A mass diagram with negative slope represents a fill operation (e.g., between station 14+75 to station 17+00), and a mass diagram with positive slope (e.g., station 11+75 to station 14+75) represents a cut operation.

B) The total grading operation is balanced, since the mass diagram is shown ending at the balance line (i.e., 0 yd³). If the mass diagram ends above the balance line, there is excess soil resulting from the project. If the mass diagram ends below the balance line, there is shortage of soil resulting from the project.

C) The section between Station 11+00 and 16+00 does NOT represent the "free-haul distance". The free-haul distance means a distance over which hauling material involves no extra cost, and is project specific. More information is required to determine the free-haul distance.

D) The section at Station 17+00 represents a transition between fill and cut. A low-point on a mass diagram represents a transition between fill and cut for the original grade. A high-point on a mass diagram represents a transition between cut and fill for the original grade.

Correct Answer: (B)

SOLUTION 29

Which of the following is false regarding temporary erosion control at a construction site? Choose the best answer.

A) In general, sand bags should be placed where they can divert and slow water and sediment flow to accumulate into predetermined deposit locations.

B) Silt fences, straw bale barriers, and sand bag barriers are all considered temporary erosion control.

C) Temporary erosion control is mandatory only at construction projects that require below-grade and slab-on-grade concrete operations.

D) The desired objective when developing a temporary erosion control plan is to keep soil at its original location.

1. What type of problem is it?

   ■ Construction: Site Development

2. What is the problem asking for?

   ■ Best answer for false statement regarding temporary soil erosion and sediment control

3. Is there extraneous information?

   ■ No

4. What references or equations are needed?

   ■ Reference construction regulations for temporary soil erosion and sediment control

5. Solution:

   Temporary erosion control is mandatory at all construction projects.

   Correct Answer: (C)

SOLUTION 30

A saturated soil sample has a dry unit weight (Υ_d) of 165 lb/ft³ and a water content (ω) of 12.5%. The sample is borrow soil to be used for a cantilever wall foundation. Find the void ratio (e) of the soil sample.

1. What type of problem is it?

   - Geotechnical: Materials

2. What is the problem asking for?

   - Void ratio (e) of the soil sample

3. Is there extraneous information?

   - The fact that it is borrow soil is not needed

   - The fact that it is a cantilever wall is not needed

4. What references or equations are needed?

   - Soil indexing formulas

   - $V_{water} = \left(\dfrac{W_{water}}{\Upsilon_{water}} \right)$

   - $e = \left(\dfrac{V_{voids}}{V_{solids}} \right)$

   - $W_{water} = \left(\omega \right)\left(W_{solids} \right)$

   - $W_{solids} = \left(\Upsilon_d \right)\left(V_{solids} \right)$

> ⊣ **"QUICK TIPS"** ├
> Always start with an equation you're trying to find, then determine all the unknowns.

5. Solution:

STEP 1 The volume of water equals the volume of voids, since the soil sample is saturated: $\left(V_{water}\right) = \left(V_{voids}\right)$

and, $V_{water} = \left(\dfrac{W_{water}}{\Upsilon_{water}}\right)$ (see above)

$$\therefore e = \left(\dfrac{V_{voids}}{V_{solids}}\right) = \left(\dfrac{V_{water}}{V_{solids}}\right)$$

STEP 2 $W_{solids} = \left(165\,\dfrac{lb}{ft^3}\right)\left(V_{solids}\right)$

$$W_{water} = (0.125)(165)\left(V_{solids}\right) = (20.625)\left(V_{solids}\right)$$

$$V_{water} = \dfrac{(20.625)\left(V_{solids}\right)}{62.4}$$

$$\therefore V_{water} = (0.3305)V_{solids}$$

STEP 3 $e = \dfrac{V_{water}}{V_s} = 0.3305$

Correct Answer: (B)

SOLUTION

A 12 ft wide rectangular channel produces a hydraulic jump from supercritical flow to subcritical flow. The depth at the point where the hydraulic jump occurs is 1.5 ft. The supercritical flow rate is 185 ft³/sec. Find the velocity of the water after the hydraulic jump.

1. What type of problem is it?

    ▪ Water Resources and Environmental: Hydraulics and Hydrology

2. What is the problem asking for?

    ▪ Velocity of the water after the hydraulic jump

3. Is there extraneous information?

    ▪ None

4. What references or equations are needed?

    ▪ Reference open channel flow

    ▪ Reference hydraulic jumps

    $$V = \frac{Q}{A}$$

    $$d_2 = -\frac{1}{2}d_1 + \sqrt{\frac{2V_1^2 d_1}{g} + \frac{d_1^2}{4}}$$

5. Solution:

    STEP 1 Convert supercritical flow to velocity (before hydraulic jump):

    $$V = \frac{Q}{A} = \frac{185 \ ft^3/sec}{(12 \ ft)(1.5 \ ft)} = 10.28 \ ft/s$$

STEP 2 Find the depth of water after the jump:

$$d_2 = -\frac{1}{2}d_1 + \sqrt{\frac{2V_1^2 d_1}{g} + \frac{d_1^2}{4}}$$

$$\therefore\ d_2 = -\frac{1}{2}(1.5\ ft) + \sqrt{\frac{2(10.28\ ft/s)^2(1.5\ ft)}{32.2\ ft/s^2} + \frac{(1.5\ ft)^2}{4}} = 2.48\ ft$$

STEP 3 Find the velocity after the jump:

$$V_2 = \frac{Q}{A_2} = \frac{185\ ft^3/sec}{(12\ ft)(2.48\ ft)} = 6.22\ ft/sec$$

Correct Answer: (D)

"THINGS TO REMEMBER"
Flow rate is the same before and after hydraulic jumps

SOLUTION 32

A 15 ft hollow beam weighs 1,575 lbs, has a pinned connection at one end, a roller at the other end, and the cross-section shown below. The statical moment of area about the neutral axis (Q) is = 76.0 in³. Find the moment of inertia (I).

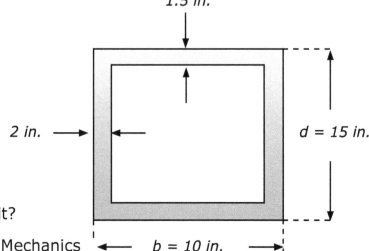

1. What type of problem is it?
   - Structural: Structural Mechanics

2. What is the problem asking for?
   - Moment of inertia, $I = \dfrac{bd^3}{12}$

3. Is there extraneous information?
   - Weight of beam not needed
   - Loading not needed
   - Connections not needed
   - Q not needed

4. What references or equations are needed?
   - Structural design

5. Solution:

   STEP 1 Find the Moment of Inertia:

   $$I = \frac{bd^3}{12} - \frac{b_1 d_1^3}{12} = \frac{(10)(15)^3}{12} - \frac{(6)(12)^3}{12} = 1948.5 \; in^4$$

   Correct Answer: (C)

SOLUTION 33

Which of the following is false regarding hydrology for stormwater runoff? Choose the best answer.

A) The "Horton Method" and "Green–Ampt Method" are methods for estimating ground water infiltration rates.

B) The Runoff Curve Number (or Curve Number, CN) is used in determining the approximate amount of direct runoff or ground water infiltration from a rainfall event.

C) A detention pond and a retention pond are the same type of runoff basin.

D) The Time of Concentration of a watershed is the time needed for water to flow from the most remote point in a watershed to the watershed outlet.

1. What type of problem is it?

- Water resources and Environmental: Hydraulics and Hydrology

2. What is the problem asking for?

- The false statement regarding stormwater runoff

3. Is there extraneous information?

- No

4. What references or equations are needed?

- Reference water infiltration rates

- Reference urban stormwater drainage

5. Solution:

Detention ponds (or Dry Basins) remain dry except during or after runoff from a rain event or snow melt. Detention ponds are primarily used for flood protection to slow down water flow and hold it for a short period of time (e.g., 24 hours) until it can be redispersed, evaporated, and/or absorbed. Detention ponds typically have an outlet at the bottom of the basin so that all of the water eventually drains out.

Retention ponds (or Wet Basins) are characterized by having a permanent pool of water that will fluctuate in response to water runoff. Retention ponds are also used for flood protection, but also for recharging the groundwater, aesthetic enhancement, and water quality improvement. Retention ponds typically have a riser outlet higher than the typical water level so that it retains a permanent pool of water.

Correct Answer: (C)

"QUICK TIPS"

If there is only one false answer, then all the other answers are true, and may be useful on the exam.

SOLUTION 34

An 18 ft Douglas Fir timber beam has a pinned connection at one end (*at A*) and a roller at the other end (*at B*), as shown below. The beam is subjected to a uniform load (*W*) of 150 lb/ft and a point load (*P*) of 1.2 kips at midspan. The beam is 4 in. wide and 12 in. deep. Find the maximum flexural stress (f_b), in ksi.

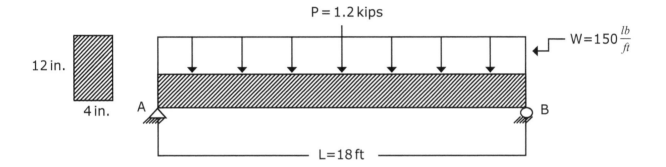

1. What type of problem is it?

   ■ Structural: Structural Mechanics

2. What is the problem asking for?

   ■ Maximum flexural stress (f_b)

3. Is there extraneous information?

   ■ The fact that the beam is made from Douglas Fir is not needed

4. What references or equations are needed?

   ■ Timber beam design

   ■ The following $f_b = \dfrac{MC}{I} = \dfrac{M_{max}}{S}$

   ■ The following equation is used for max bending moment for rectangular beam

   $$M_{max} = \frac{WL^2}{8} + \frac{PL}{4}$$

5. Solution:

STEP 1

$$M = \frac{WL^2}{8} + \frac{PL}{4} = \frac{\left(150 \frac{lb}{ft}\right)(18\ ft)^2}{8} + \frac{(1,200\ lbs)(18\ ft)}{4} = 11,475\ ft\text{-}lbs$$

STEP 2

$$C = h/2 = 6\ in.$$

STEP 3

$$I = \frac{bh^3}{12} = \frac{(4\ in.)(12\ in.)^3}{12} = 576\ in.^4$$

STEP 4

$$f_b = \frac{11,475\ ft\text{-}lb\left(\frac{12\ in.}{1ft}\right)(6\ in.)}{576\ in.^4} = 1,435\ psi$$

Correct Answer: (B)

SOLUTION

35

A welded and seamless steel pipe with 33 in. circumference provides water from a large water tank on a hill above several hotels near Yosemite National Park in California. The water tank is 700 ft above the subdivision. The pipe system is approximately 3/4 mile away. The flow inside the pipe is determined to be 3.5 ft³/sec. Find the head loss due to friction (h_f) of the pipe between the tank and the subdivision, in feet.

1. What type of problem is it?

   ■ Water Resources and Environmental: Hydraulics and Hydrology

2. What is the problem asking for?

   ■ Head loss due to friction of the pipe between the tank and the subdivision

3. Is there extraneous information?

   ■ The evaluation of the tank above the subdivision is not needed

   ■ The fact that the pipe is for hotels on in Yosemite National park is not needed

4. What references or equations are needed?

   ■ Reference fluid dynamics, energy loss due to friction for turbulent flow, and Hazen—Williams equation:

$$h_{f,feet} = \frac{4.72\left(L_{ft}\right)\left(Q_{cfs}\right)^{1.85}}{\left(C^{1.85}\right)\left(D_{ft}\right)^{4.87}} \ or \ h_{f,feet} = \frac{10.44\left(L_{ft}\right)\left(Q_{gpm}\right)^{1.85}}{\left(C^{1.85}\right)\left(D \ in.\right)^{4.87}}$$

5. Solution:

STEP 1 Find the Hazen—Williams Roughness coefficient (C) for welded and seamless steel pipe using general tables:

- C = 100

STEP 2 Solve for diameter (D) of the pipe:

$$\pi D = 33 \ in.$$

$$\therefore \ D = \frac{33 \ in.}{\pi} = 10.5 \ in. = 0.875 \ ft$$

STEP 3 Calculate the head loss, in feet $h_{f,feet}$:

$$h_{f,feet} = \frac{(4.72)(3960 \ ft)\left(3.5 \frac{ft^3}{sec}\right)^{1.85}}{(100)^{1.85}(0.875 \ ft)^{4.87}} = 72.5 \ ft$$

Correct Answer: (B)

SOLUTION

36

A truss system has a pinned connection on one side (*at Point A*) and a roller on the other side (*at Point H*), as shown below. 2 kips acts downward at Point C and 5 kips acts downward at Point E and Point F. Neglect dead load of the framing members. Find the force in member BE.

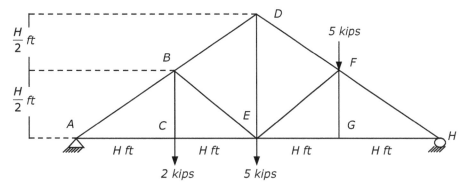

---

1. What type of problem is it?

   ■ Structural: Structural Mechanics

2. What is the problem asking for?

   ■ The question is asking for the force in member BE

3. Is there extraneous information?

   ■ No

4. What references or equations are needed?

   ■ Reference forces in axial members

   ■ Reference forces in determinant trusses

   ■ Reference Free-Body Diagrams

   ■ Reference Method of Joints, Method of Sections, and Cut-and Sum Method

5. Solution:

 STEP 1 Find reactions on the truss system at Point H by taking the sum of moments about Point A.

$$\sum M_A = 0; (5\ kips)(3H) + (5\ kips)(2H) + (2\ kips)(H) = (Y_H)(4H)$$

$$\therefore\ (27\ kips)(H) = (Y_H)(4H)$$

$$\therefore\ Y_H = \frac{27\ kips}{4} = 6.75\ kips$$

 STEP 2 Find reactions on the truss system at Point A by taking the sum of forces in the x-direction and y-direction.

$$\sum F_{x-direction} = 0$$

$$\sum F_{x-direction} = 0;\ Y_A + Y_H = 2\ kips + 5\ kips + 5\ kips;$$

where $Y_H = 6.75\ kips$

$$\therefore\ Y_A = 2\ kips + 5\ kips + 5\ kips - 6.75\ kips = 5.25\ kips$$

> ┌ **"THINGS TO THINK ABOUT"** ┐
> After finding the external forces, can you use the Method of Sections to find the force in BE, or will there be too many unknowns?

 STEP 3 Find reactions in member AB and member AC by using Method of Joints at Junction A.

$$\theta = tan^{-1}\left(\frac{H/2}{H}\right) = 26.56°$$

> ┌ **"THINGS TO REMEMBER"** ┐
> Can you eliminate 2 of the answers from the multiple choice questions?

$$\sum F_{y-direction} = 0;\ 5.25\ kips + AB(sin(26.56°))$$

$$\sum F_{x-direction} = 0$$

$$\therefore\ AB = \frac{-5.25\ kips}{sin(26.56°)} = 11.74\ kips;$$

$$AC + AB(cos(26.56°)) = 0$$

$$\therefore\ AC = -AB(cos(26.56°)) = -11.74\ kips(cos(26.56°)) = -10.5\ kips$$

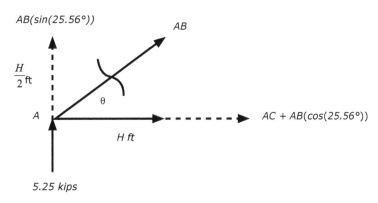

STEP 4 Find reactions in member BC and member CE by using Method of Joints at Junction C.

$$\sum F_{y-direction} = 0; \ AC = CE = -10.5 \ kips$$

$$\sum F_{x-direction} = 0; \ By \ observation, \ BC = 2 \ kips$$

> **"THINGS TO THINK ABOUT"**
> What if you were asked if member AB was in tension or compression?

STEP 5 Find reactions in member BE by using Method of Joints or Method of Sections (2 equations, 2 unknowns either way).

By Method of Joints at Junction B:

> **"THINGS TO THINK ABOUT"**
> Can you solve using Method of Sections?

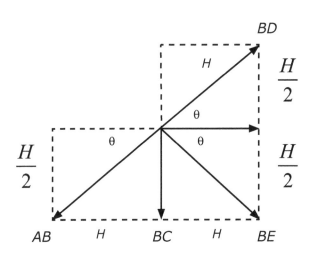

which gives:

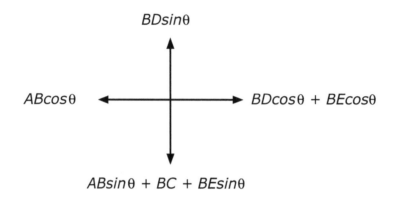

$\therefore BD\sin\theta = AB\sin\theta + BC + BE\sin\theta,$

and $BD\cos\theta + BE\cos\theta = AB\cos\theta$

$\therefore BD\sin(26.56°) = (-11.74\ kips)(\sin(26.56°)) + 2\ kips + BE\sin(26.56°)$

$\therefore BD = -7.26\ kips + BE$ [equation 1]

Since: $BD = AB - BE$

$\therefore BD = -11.74 - BE$ [equation 2]

By combining equation 1 and 2:

$-11.74 - BE = -7.26 + BE$

By solving for BE:

$BE = -2.24\ kips$

$\therefore BE = 2.24\ kips$ [compression]

Correct Answer: (C)

SOLUTION 37

Which of the following is false regarding soil stress at a construction site? Choose the best answer.

A) Effective stress in soils is the difference between the total stress and the pore water pressure.

B) Above the water table, total and effective stresses are equal.

C) Below the water table, the pore pressure is a product of the specific weight of water and the height of the water above the point measured.

D) In dry soil, particles at a point underground experience a total overhead stress (depth underground multiplied by the specific weight of the soil) that is less than the stress at the same point in saturated soil.

1. What type of problem is it?

   ■ Geotechnical: Soil Mechanics

2. What is the problem asking for?

   ■ False statement regarding soil stress at a construction project

3. Is there extraneous information?

   ■ No

4. What references or equations are needed?

   ■ Reference geotechnical engineering, soil stress, and total and effective stresses

5. Solution:

   Particles underground in dry soil experience more total overhead stress than particles in saturated soil, because of the buoyancy effects of water. The total stress below the surface is decreased by the product of the height of the water table and the specific weight of water. This concept is known as the effective stress of the soil, equal to the difference in a soil's total stress and pore water pressure.

Correct Answer: (D)

SOLUTION

38

Timber shoring and formwork support a 6 in. concrete slab (*150 pcf*), as shown below. The concrete will be cast-in-place over 3/4 in. plywood sheathing (*2.3 psf*). The 6 x 6 shores are spaced @ 48 in. o.c. The 4 x 6 stringers are spaced @ 5 ft o.c. The 2 x 4 joists (*1.5 in. x 3.5 in. actual*) are spaced at 16 in. o.c. The live load will be 20 psf. Assume the dead load is only the concrete and the plywood. Some lateral cross-bracing is installed for the shores. Metal bracket connections mechanically attach the shores to sill plates. Find the maximum bending stress (σ_b) on the joists.

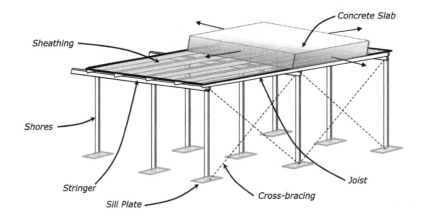

1. What type of problem is it?

   ■ Construction: Means and Methods

2. What is the problem asking for?

   ■ The question is asking for the maximum bending stress (σ_b) on the joists

3. Is there extraneous information?

   ■ The size of the stringers are not needed to answer the question

   ■ The size of the shores are not needed to answer the question

> **"THINGS TO THINK ABOUT"**
> What if you were asked for maximum tributary forces on a perimeter stringer?

- Solution does not require information regarding the lateral cross-bracing that is installed for the shores

- Solution does not require information regarding the metal bracket connections that mechanically attach the shores to sill plates

4. What references or equations are needed?

- Reference Timber design

- Reference Shoring and Formwork design

- The following equation is used for maximum bending stress for the maximum bending moment (M_{max}) divided by the section modulus (S):

$$\sigma_b = \frac{M_{max}c}{1} = \frac{M_{max}}{S}$$

- The following equation is used for maximum bending moment (M_{max}) for a simply supported span (between stringer supports):

$$M_{max} = \frac{wl^2}{8}$$

- The following equation is used for the moment of inertia (I) for a rectangular section (e.g., the joist):

$$I = \frac{bh^3}{12}$$

- The following equation is used for the distance from the neutral axis to the outer surface where the maximum stress occurs (c):

$$c = \frac{h}{2}$$

- The section modulus (S) is the moment of inertia (I) divided by the distance from the neutral axis to the outer surface where the maximum stress occurs (c):

$$S = \frac{I}{c}$$

■ The following equation is used for the section modulus (S) for a rectangular section (e.g., the joist):

$$S = \frac{\frac{bh^3}{12}}{\frac{h}{2}} = \frac{bh^2}{6}$$

5. Solution:

STEP 1 Find the total live and dead loads:

Live load is 20 psf.

$$Dead\ Load = \left(150\ pcf \times \frac{6\ in.}{12\frac{in.}{ft}}\right) = 75\ psf$$

Total Load = 20 psf + 75 psf + 2.3 psf = 97.3 psf

STEP 2 Find the maximum tributary area on the joists (note: use inner joists where the tributary loads are the greatest):

$$\therefore\ 97.3\ psf \times \left(16\ in. \times \frac{1\ ft}{12\ in.}\right) = 129.74\ plf$$

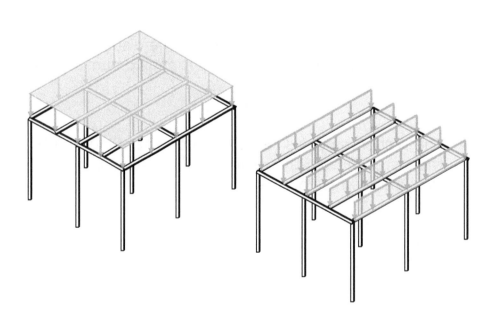

STEP 3 Find the maximum bending moment on the joist, (M_{max}):

$$M_{max} = \frac{(129.74\ plf)(5\ ft)^2}{8} = 405.52\ lb\text{-}ft$$

STEP 4 Find the distance from the section modulus (S) for a rectangular section (e.g., the joist):

$$S = \frac{\frac{bh^3}{12}}{\frac{h}{2}} = \frac{bh^2}{6} = \frac{(1.5\ in.)(3.5\ in.)^2}{6} = 3.0625\ in.^3$$

STEP 5 Find maximum bending stress:

$$\sigma_b = \frac{M_{max}c}{I} = \frac{M_{max}}{S} = \frac{(405.52\ lb\text{-}ft)\left(\dfrac{12\ in.}{1\ ft}\right)}{3.0625\ in^3} = 1,588.98\ psi$$

Correct Answer: (A)

SOLUTION

39

A network of pipes is configured in parallel under a playground in San Diego, California, as shown. Two of the pipes are schedule–40 steel and one of the pipes is bituminous-lined cast iron. The nominal sizes of the pipes are 3 in., 4 in., and 6 in., with Hazen–Williams loss coefficients of 100, 100, and 140, as shown. The 3 in. pipe is 150 ft long. The 4 in. pipe is 90 ft long. The 6 in. pipe is 225 ft long. Minor losses are insignificant. Water enters the pipe system at A at 4.2 ft³/sec. Find the total friction loss between junctions A and B.

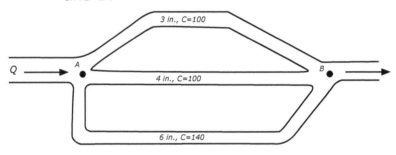

1. What type of problem is it?

   - Water Resources and Environmental: Hydraulics and Hydrology

   ┌─ **"THINGS TO REMEMBER"** ─┐
   Whenever C is in the problem statement, reference Hazen–Williams or Chezy

2. What is the problem asking for?

   - Total friction loss between junctions A and B

3. Is there extraneous information?

   - The fact that the pipes are under a playground is not needed

   - The fact that the playground is in San Diego is not needed

   - The material type is not needed since the Hazen–Williams friction loss coefficients are given

4. What references or equations are needed?

   - Reference Parallel pipe system

   - Hazen–Williams friction loss $\quad h_f = \dfrac{3.022\ V^{1.85}L}{C^{1.85}D^{1.165}}$

- $Q = VA$

- For three Parallel pipes

  $Q_{total} = Q_1 + Q_2 + Q_3$

5. Solution:

STEP 1 Find velocity (V) for each pipe by using Hazen–Williams Equation.

Solving for V: $V = \dfrac{0.550\, D^{0.63} h_f^{0.54} C}{L^{0.54}}$

$$V_{3in.} = \frac{(0.55)\left(\dfrac{3\ in.}{12}\right)^{0.63} h_f^{0.54}(100)}{(150)^{0.54}} = (1.53)h_f^{0.54}$$

$$V_{4in.} = \frac{(0.55)\left(\dfrac{4\ in.}{12}\right)^{0.63} h_f^{0.54}(100)}{(90)^{0.54}} = (2.41)h_f^{0.54}$$

$$V_{6in.} = \frac{(0.55)\left(\dfrac{6\ in.}{12}\right)^{0.63} h_f^{0.54}(140)}{(225)^{0.54}} = (2.67)h_f^{0.54}$$

STEP 2 Find Q for each pipe by multiplying the velocity (V) by the pipe area (A):

$$Q_{3in.} = (1.53)h_f^{0.54}(A_{3in.}) = (1.53)h_f^{0.54}\left[\frac{\pi}{4}\left(\frac{3\ in.}{12}\right)^2\right] = 0.075 h_f^{0.54}$$

$$Q_{4in.} = (2.41)h_f^{0.54}(A_{4in.}) = (2.41)h_f^{0.54}\left[\frac{\pi}{4}\left(\frac{4\ in.}{12}\right)^2\right] = 0.210 h_f^{0.54}$$

$$Q_{6in.} = (2.67)h_f^{0.54}(A_{6in.}) = (2.67)h_f^{0.54}\left[\frac{\pi}{4}\left(\frac{6\ in.}{12}\right)^2\right] = 0.534 h_f^{0.54}$$

STEP 3 For parallel pipes, the total flow is the sum of the flow rates in each branch and h_f is the same:

$Q_{total} = Q_{3in} + Q_{4in} + Q_{6in}$

$\therefore\ 4.2\,\dfrac{ft^3}{sec} = 0.075 h_f^{0.54} + 0.210 h_f^{0.54} + 0.534 h_f^{0.54}$

$\therefore\ h_f = 20.64\ ft$ **Correct Answer: (B)**

SOLUTION

A 2,700 lb steam hammer is dropped from a height of 3.7 ft into soil with unit weight (Υ_{soil}) of 110 lb/ft³. The water content (ω) is 8% and the specific gravity of the solids (SG_{solids}) is 2.67. The pile has driven 1.2 ft in the last 10 blows. Find the allowable capacity of the pile (Q_a) using the Engineering News (ENR) Formula if the driven weight is 2,995 lbs.

1. What type of problem is it?

   ▪ Geotechnical: Soil mechanics

2. What is the problem asking for?

   ▪ Find the allowable capacity of the pile using the Engineering News Record (ENR) formula

3. Is there extraneous information?

   ▪ The unit weight of the soil (Υ_{soil}) is not needed

   ▪ The water content (ω) is not needed

   ▪ The specific gravity of the solids (SG_{solids}) is not needed

4. What references or equations are needed?

   ▪ ENG formula for pile capacity:

   $$Q_a = \frac{2\,W_{hammer}H}{S + (0.1)\left(\dfrac{W\ driven}{W_{hammer}}\right)}$$

5. Solution:

   STEP 1 Find S: $S = \dfrac{1.2\ ft}{10}\left(\dfrac{12\ in.}{1\ ft}\right) = 1.44\ in.$

   STEP 2 Find Q_a: $Q_a = \dfrac{2(2,700\ lbs)(3.7\ ft)}{1.44\ in. + (0.1)\left(\dfrac{2,995\ lbs}{2,700\ lbs}\right)}$

   $\therefore Q_a = 12.9\ kips$

   Correct Answer: (C)

DISCLAIMER AND TERMS OF USE AGREEMENT

The information contained in this guide is the opinion of the individual authors based on their personal observations and years of experience. Neither the authors nor publisher assume any liability whatsoever for the use of or inability to use any or all information contained in this publication. Use this information at your own risk.

The author and publisher of this guide and the accompanying materials have used their best efforts in preparing this guide. The author and publisher make no representation or warranties with respect to the accuracy, applicability, competency, or completeness of the contents of this guide. The information contained in this guide is strictly for educational purposes. Therefore, if you wish to apply ideas contained in this guide, you are taking full responsibility for your actions.

EVERY EFFORT HAS BEEN MADE TO ACCURATELY REPRESENT THIS PRODUCT AND IT'S POTENTIAL. EXAMPLES IN THESE MATERIALS ARE NOT TO BE INTERPRETED AS A PROMISE OR GUARANTEE OF PASSING THE PROFESSIONAL ENGINEERING EXAM. SUCCESS IS ENTIRELY DEPENDENT ON THE PERSON USING THE PRODUCT, IDEAS, AND TECHNIQUES. YOUR LEVEL OF SUCCESS IN ATTAINING THE RESULTS YOU DESIRE, DEPENDS ON THE TIME YOU DEVOTE TO THE EFFORT, IDEAS AND TECHNIQUES MENTIONED, KNOWLEDGE, AND VARIOUS SKILLS. SINCE THESE FACTORS DIFFER ACCORDING TO INDIVIDUALS, WE CANNOT GUARANTEE YOUR SUCCESS. NOR ARE WE RESPONSIBLE FOR ANY OF YOUR ACTIONS. MANY FACTORS WILL BE IMPORTANT IN DETERMINING YOUR ACTUAL RESULTS AND NO GUARANTEES ARE MADE THAT YOU WILL ACHIEVE RESULTS YOU WISH FOR.

The author and publisher disclaim any warranties (express or implied), merchantability, or fitness for any particular purpose. The author and publisher shall in no event be held liable to any party for any direct, indirect, punitive, special, incidental or other consequential damages arising directly or indirectly from any use of this material, which is provided "as is", and without warranties.

The author and publisher do not warrant the performance, effectiveness or applicability of any sites listed or linked to in this guide. All links are for information purposes only and are not warranted for content, accuracy or any other implied or explicit purpose.

Errata

passthecivilPE is very thankful to readers who notify us of any possible errors and/or omissions. Your positive feedback and constructive criticism allows us to improve the quality and provide better use of your study time.

You can report errors and/or omissions and request any corrections and addendums at passthecivilPE.com. Thank you for your continued support.

PasstheCivilPE Exam Checklist

Before the Exam:

1. Check with your State Board for a Civil Professional Engineering (P.E.) exam Application and current requirements for the exams (check the due dates for your state and for NCEES – they likely have different due dates).

2. Submit all required information to your State Board and NCEES.

3. Develop a study plan for the exam. Consider tackling your depth subject first, several months in advance. This will allow you to judge about how long it will take for the other subjects so you can develop your study plan. Make sure you come back and study your depth again a couple weeks before the test.

4. Notify your employer that you will be taking the exam.

5. Study for the exam, form a study group, take a review course, take several timed practice exams (this is most important), etc.

6. Familiarize yourself with your calculator and reference materials by using them during your study.

7. Wear earplugs while you're studying if you plan to use them during the exam. Earplugs can be uncomfortable and distracting after wearing them for hours.

8. Get lots of rest a few days before the exam. If you are flying into the exam, do it a day early (at least).

9. Know where to go for the exam. These tests are usually held in giant complexes with several different buildings and parking lots. It won't help you to get lost the day of the exam.

Morning of the Exam:

1. Bring your NCEES "exam day" information and admission notice to the exam (this will be available a few weeks before the exam on the NCEES website).

2. Bring your photo ID (check the Exam Day Policies for appropriate ID).

3. Bring your Approved Calculator.

4. Turn your phone off and leave it in your car. Your phone isn't allowed in the exam room and it will be one less thing you need to worry about.

5. Bring ear plugs.

6. Wear something comfortable and dress in layers. Bring a light sweater.

7. Know where you're going. Get to the exam site early, find a quiet area, and review your notes. Resist the urge to talk with others, unless you think this will relieve some stress.

8. Be prepared for an hour or more after the doors to the exam site open before taking the first 4-hour test. There will be a lot of paperwork to fill out and rules the proctors will read to you.

9. Bring ALL of your references.

10. Bring a packable, bag lunch. There will likely be no food at the exam.

During the Exam:

Go to the passthecivilPE Exam Advice page and read the compiled list of information and advice to use during the exam.

Name: _____

Student ID: _____

Date: _____

Test #: _____

SCORE:

CHOOSE THE BEST ANSWER.
FILL-IN BUBBLE COMPLETELY: Ⓐ Ⓑ ● Ⓓ

1.	Ⓐ Ⓑ Ⓒ Ⓓ	11.	Ⓐ Ⓑ Ⓒ Ⓓ	21.	Ⓐ Ⓑ Ⓒ Ⓓ	31.	Ⓐ Ⓑ Ⓒ Ⓓ
2.	Ⓐ Ⓑ Ⓒ Ⓓ	12.	Ⓐ Ⓑ Ⓒ Ⓓ	22.	Ⓐ Ⓑ Ⓒ Ⓓ	32.	Ⓐ Ⓑ Ⓒ Ⓓ
3.	Ⓐ Ⓑ Ⓒ Ⓓ	13.	Ⓐ Ⓑ Ⓒ Ⓓ	23.	Ⓐ Ⓑ Ⓒ Ⓓ	33.	Ⓐ Ⓑ Ⓒ Ⓓ
4.	Ⓐ Ⓑ Ⓒ Ⓓ	14.	Ⓐ Ⓑ Ⓒ Ⓓ	24.	Ⓐ Ⓑ Ⓒ Ⓓ	34.	Ⓐ Ⓑ Ⓒ Ⓓ
5.	Ⓐ Ⓑ Ⓒ Ⓓ	15.	Ⓐ Ⓑ Ⓒ Ⓓ	25.	Ⓐ Ⓑ Ⓒ Ⓓ	35.	Ⓐ Ⓑ Ⓒ Ⓓ
6.	Ⓐ Ⓑ Ⓒ Ⓓ	16.	Ⓐ Ⓑ Ⓒ Ⓓ	26.	Ⓐ Ⓑ Ⓒ Ⓓ	36.	Ⓐ Ⓑ Ⓒ Ⓓ
7.	Ⓐ Ⓑ Ⓒ Ⓓ	17.	Ⓐ Ⓑ Ⓒ Ⓓ	27.	Ⓐ Ⓑ Ⓒ Ⓓ	37.	Ⓐ Ⓑ Ⓒ Ⓓ
8.	Ⓐ Ⓑ Ⓒ Ⓓ	18.	Ⓐ Ⓑ Ⓒ Ⓓ	28.	Ⓐ Ⓑ Ⓒ Ⓓ	38.	Ⓐ Ⓑ Ⓒ Ⓓ
9.	Ⓐ Ⓑ Ⓒ Ⓓ	19.	Ⓐ Ⓑ Ⓒ Ⓓ	29.	Ⓐ Ⓑ Ⓒ Ⓓ	39.	Ⓐ Ⓑ Ⓒ Ⓓ
10.	Ⓐ Ⓑ Ⓒ Ⓓ	20.	Ⓐ Ⓑ Ⓒ Ⓓ	30.	Ⓐ Ⓑ Ⓒ Ⓓ	40.	Ⓐ Ⓑ Ⓒ Ⓓ

Name: ....

Student ID: ....

Date: ....

Test #: ....

SCORE:

CHOOSE THE BEST ANSWER.
FILL-IN BUBBLE COMPLETELY: Ⓐ Ⓑ Ⓒ Ⓓ

1. Ⓐ Ⓑ Ⓒ Ⓓ	11. Ⓐ Ⓑ Ⓒ Ⓓ	21. Ⓐ Ⓑ Ⓒ Ⓓ	31. Ⓐ Ⓑ Ⓒ Ⓓ
2. Ⓐ Ⓑ Ⓒ Ⓓ	12. Ⓐ Ⓑ Ⓒ Ⓓ	22. Ⓐ Ⓑ Ⓒ Ⓓ	32. Ⓐ Ⓑ Ⓒ Ⓓ
3. Ⓐ Ⓑ Ⓒ Ⓓ	13. Ⓐ Ⓑ Ⓒ Ⓓ	23. Ⓐ Ⓑ Ⓒ Ⓓ	33. Ⓐ Ⓑ Ⓒ Ⓓ
4. Ⓐ Ⓑ Ⓒ Ⓓ	14. Ⓐ Ⓑ Ⓒ Ⓓ	24. Ⓐ Ⓑ Ⓒ Ⓓ	34. Ⓐ Ⓑ Ⓒ Ⓓ
5. Ⓐ Ⓑ Ⓒ Ⓓ	15. Ⓐ Ⓑ Ⓒ Ⓓ	25. Ⓐ Ⓑ Ⓒ Ⓓ	35. Ⓐ Ⓑ Ⓒ Ⓓ
6. Ⓐ Ⓑ Ⓒ Ⓓ	16. Ⓐ Ⓑ Ⓒ Ⓓ	26. Ⓐ Ⓑ Ⓒ Ⓓ	36. Ⓐ Ⓑ Ⓒ Ⓓ
7. Ⓐ Ⓑ Ⓒ Ⓓ	17. Ⓐ Ⓑ Ⓒ Ⓓ	27. Ⓐ Ⓑ Ⓒ Ⓓ	37. Ⓐ Ⓑ Ⓒ Ⓓ
8. Ⓐ Ⓑ Ⓒ Ⓓ	18. Ⓐ Ⓑ Ⓒ Ⓓ	28. Ⓐ Ⓑ Ⓒ Ⓓ	38. Ⓐ Ⓑ Ⓒ Ⓓ
9. Ⓐ Ⓑ Ⓒ Ⓓ	19. Ⓐ Ⓑ Ⓒ Ⓓ	29. Ⓐ Ⓑ Ⓒ Ⓓ	39. Ⓐ Ⓑ Ⓒ Ⓓ
10. Ⓐ Ⓑ Ⓒ Ⓓ	20. Ⓐ Ⓑ Ⓒ Ⓓ	30. Ⓐ Ⓑ Ⓒ Ⓓ	40. Ⓐ Ⓑ Ⓒ Ⓓ

Name: _____

Student ID: _____

Date: _____

Test #: _____

SCORE:

CHOOSE THE BEST ANSWER.
FILL-IN BUBBLE COMPLETELY: Ⓐ Ⓑ ⬤ Ⓓ

1. Ⓐ Ⓑ Ⓒ Ⓓ	11. Ⓐ Ⓑ Ⓒ Ⓓ	21. Ⓐ Ⓑ Ⓒ Ⓓ	31. Ⓐ Ⓑ Ⓒ Ⓓ
2. Ⓐ Ⓑ Ⓒ Ⓓ	12. Ⓐ Ⓑ Ⓒ Ⓓ	22. Ⓐ Ⓑ Ⓒ Ⓓ	32. Ⓐ Ⓑ Ⓒ Ⓓ
3. Ⓐ Ⓑ Ⓒ Ⓓ	13. Ⓐ Ⓑ Ⓒ Ⓓ	23. Ⓐ Ⓑ Ⓒ Ⓓ	33. Ⓐ Ⓑ Ⓒ Ⓓ
4. Ⓐ Ⓑ Ⓒ Ⓓ	14. Ⓐ Ⓑ Ⓒ Ⓓ	24. Ⓐ Ⓑ Ⓒ Ⓓ	34. Ⓐ Ⓑ Ⓒ Ⓓ
5. Ⓐ Ⓑ Ⓒ Ⓓ	15. Ⓐ Ⓑ Ⓒ Ⓓ	25. Ⓐ Ⓑ Ⓒ Ⓓ	35. Ⓐ Ⓑ Ⓒ Ⓓ
6. Ⓐ Ⓑ Ⓒ Ⓓ	16. Ⓐ Ⓑ Ⓒ Ⓓ	26. Ⓐ Ⓑ Ⓒ Ⓓ	36. Ⓐ Ⓑ Ⓒ Ⓓ
7. Ⓐ Ⓑ Ⓒ Ⓓ	17. Ⓐ Ⓑ Ⓒ Ⓓ	27. Ⓐ Ⓑ Ⓒ Ⓓ	37. Ⓐ Ⓑ Ⓒ Ⓓ
8. Ⓐ Ⓑ Ⓒ Ⓓ	18. Ⓐ Ⓑ Ⓒ Ⓓ	28. Ⓐ Ⓑ Ⓒ Ⓓ	38. Ⓐ Ⓑ Ⓒ Ⓓ
9. Ⓐ Ⓑ Ⓒ Ⓓ	19. Ⓐ Ⓑ Ⓒ Ⓓ	29. Ⓐ Ⓑ Ⓒ Ⓓ	39. Ⓐ Ⓑ Ⓒ Ⓓ
10. Ⓐ Ⓑ Ⓒ Ⓓ	20. Ⓐ Ⓑ Ⓒ Ⓓ	30. Ⓐ Ⓑ Ⓒ Ⓓ	40. Ⓐ Ⓑ Ⓒ Ⓓ

This guide was developed because we know that practice is the most essential component to passing the Civil Professional Engineering (P.E.) Exam. Training with materials similar in format, timing, language, and style will increase your comfort level and prepare you for mastering the exam when it counts the most. This guide provides you with necessary information in the form of a combined practice exam and study guide that will give you confidence and prepare you for passing the Civil Professional Engineering (P.E.) Exam.

www.passthecivilPE.com

Exam topics covered include:

- ✓ Construction
- ✓ Transportation
- ✓ Structural
- ✓ Water Resources and Environmental
- ✓ Geotechnical

Including the 8 NCEES distinct categories for breath exam topics!

Project Planning, Means & Methods, Soil Mechanics, Structural Mechanics, Hydraulics & Hydrology, Geometrics, Materials, and Site Development

CPSIA information can be obtained
at www.ICGtesting.com
Printed in the USA
BVHW061547060222
628184BV00005B/161